UNDERSTANDING AND MEASURING VIBRATION

THE WYKEHAM TECHNOLOGICAL SERIES

for universities and institutes of technology

General Editor:

J. Thomson, M.A., D.Sc., F.I.E.E., F.Inst.P.
recently Chief of Technical Studies,
Sangamo Weston, Ltd.

THIS BOOK is one of a series in which authorities will attempt to introduce young people to the various technologies in the subjects discussed. Primarily these books are aimed at young graduates or apprentices who are about to begin a career in industry or who have just started work. For this reason more attention has been paid to giving a broad view of each subject than to ‘ necessary and sufficient ’ mathematical proofs. Indeed the use of mathematics has been sparing.

Another feature of the Technological Series is the mixture of disciplines. For example, when micro-circuits are being discussed, the physical basis of semi-conduction phenomena is inextricably mixed up with the chemistry of almost pure single crystal formation and with the engineering necessary to produce a monolithic circuit. This admixture of disciplines is a feature of industrial research and development. Indeed, it might be argued that this feature of industrial work makes it very unsuitable for the so-called ‘ specialist ’ who only recognises his own discipline.

It is hoped both by the authors and by the publishers that this series will open up some new horizons to the young scientist and engineer in industry.

UNDERSTANDING AND MEASURING VIBRATION

R. H. Wallace

WYKEHAM PUBLICATIONS (LONDON) LTD
A member of the Taylor & Francis Group
LONDON AND WINCHESTER
1970

First published 1970 by Wykeham Publications (London) Ltd.

Cover illustration—A flame as a gas vibration transducer, see p. 35.

ISBN 0 85109 180 6

Printed in Great Britain by Taylor & Francis, Ltd.
10–14 Macklin Street, London, WC2B 5NF

Distribution:

UNITED KINGDOM, EUROPE, MIDDLE EAST AND AFRICA
Chapman & Hall Ltd. (a member of Associated Book Publishers Ltd.), 11 New Fetter Lane, London, E.C.4 and North Way, Andover, Hampshire.

UNITED STATES OF AMERICA, CANADA AND MEXICO
Springer–Verlag New York Inc., 175 Fifth Avenue, New York, New York 10010.

AUSTRALIA AND NEW GUINEA
Hicks Smith & Sons Pty. Ltd., 301 Kent Street, Sydney, N.S.W. 2000.

NEW ZEALAND AND FIJI
Hicks Smith & Sons Ltd., 238 Wakefield Street, Wellington.

ALL OTHER TERRITORIES
Taylor & Francis Ltd., 10–14 Macklin Street, London, WC2B 5NF.

INTRODUCTION

WHILE there is no dearth of academic works on the dynamics of vibration which cover the formal field completely, the manner of their presentation is such that the practising engineer often finds them difficult to apply to his complex machines.

The purpose of the present book is to remedy the situation; and enable engineers, concerned mainly with design development and production, to visualize for themselves the manner in which whole engines, and their separate parts, vibrate in practice.

Mathematics has been almost entirely avoided—the various formulae can be found in academic textbooks when needed—and the book attempts to help the engineer to appreciate the various conditions in which harmful vibration takes place.

The young engineer just starting in industry and also the old hand not previously familiar with dynamics will find that most actual engines appear unrelated to the simple examples of elemental vibrating systems given in many textbooks. Though engines comprise assemblies of many separately vibrating systems (each simple in itself) the interaction between them can be difficult to visualize.

The book shows how vibration can be avoided or reduced, and the manner in which it makes its effects known. Often a particular method of measurement can colour the results obtained, so a large part of the work deals with measurement and the interpretation of the results.

The engineer entering the field of vibration should find the contents helpful; while the design, development and production engineer should be able to acquire a ' feel ' for what goes on in engines—as far as vibration is concerned. Engineers already well versed in vibration work may find parts of the book oversimplified, but should find the examples interesting—and perhaps occasionally illuminating.

Since the fields in which vibration is of significance are so widespread —and also because the author's preoccupation in the last decade has been with aero engines—most of the illustrations are drawn from this particular area. The observations made are, however, equally applicable in respect of heavier machinery and structures; although the emphasis as between resonant and forced vibrations, and also the nature and magnitude of exciting forces, will vary from case to case.

ACKNOWLEDGEMENTS

THE author thanks particularly Mr. D. McLean and Mr. D. R. Cooper, of Rolls-Royce Ltd., for their encouragement and advice.

Amongst the many colleagues who have helped in various ways, Mr. J. Feakes, Mr. K. Herbert, Mr. R. Graham and Mr. L. M. Smith gave valuable help in layout and checking the manuscript. Mr. C. McKay was responsible for the artist's impressions of vibrating parts and Mrs. F. Gillies and Miss D. Welsh typed the many drafts of the manuscript. The writer thanks the Board of the Aero Engine Division of Rolls-Royce for permission to publish many of the examples and photographs. Any opinions expressed are, of course, those of the author and are not necessarily endorsed by Rolls-Royce Ltd., or by the Editor or Publisher.

CONTENTS

LIST OF ILLUSTRATIONS

CHAPTER 1

explanation of vibration

THIS chapter is included for the less-experienced reader, because a long experience has shown the author that many engineers find difficulty in understanding vibration—even if they have worked out many classical examples. In other words, they see in mathematical terms what must happen but do not visualize the effects clearly. For this reason in particular this chapter proceeds in the main by examples from everyday life. In order to understand what different forms of vibration really imply, the dictionary is a good starting place. The *Oxford Dictionary* gives " vibration, n. Vibrating, oscillation; (phys) rapid motion to and fro, esp. of the parts of a fluid or an elastic solid whose equilibrium is disturbed ". In this connection it is well to remember that, whether apparent or not, all matter, gaseous, liquid or solid is capable of executing vibrations ; and, in fact, so are the elementary particles of which matter is composed.

To save confusion, frequencies of vibration throughout the book are given in Hertz (abbreviated Hz, kHz, etc.) as this nomenclature is coming into general use. Older readers may still think in cycles per second (abbreviated c.p.s., kc, etc.). One Hertz is one cycle per second.

A major distinction in this wide field is between ' forced ' vibrations of a component or engine and what is commonly called ' resonant ' vibration—the latter being always of a periodic nature while the former may or may not be. In general terms the piston of a car engine is vibrating and so is a tree shaking in a gusty wind or a boat on a rough sea! But while the piston is designed to vibrate without distress, the same may not be true of the tree or—sometimes—the boat. Resonant vibrations, as we shall see later on, are generally undesirable and may well be disastrous.

For practical engineering purposes it is better to narrow the field and it is usual to allocate the term ' vibration ' predominantly to undesired periodic motion. In music, of course, the opposite is the case—since all musical instruments use periodic vibrations to make sound, as even Pythagoras well knew. We might say that vibration in engineering is more akin to noise in acoustics ; an undesired, but to a degree, inescapable by-product of the machine, either in terms of external sound or damage within itself.

The old intuitive idea that a machine which makes more noise is in a worse state is based simply on the fact that noise generated by parts of a mechanism must imply unnecessary stresses and strains within. Again to take a parallel from everyday life, all the ways within a machine which produce undesirable vibration are met with in music, be it classical or popular. Drums are struck and strings are plucked (impulse excitation), strings are bowed (stick–slip excitation) and organ pipes and other wind instruments are blown (wind excitation). Indeed a knowledge of how instruments work is of great help in understanding mechanical vibrations in engines. While it is not easy to observe how an engine part, vibrating only a few thousandths of an inch, moves, it is easy to understand the similar vibrations of the lower strings of a guitar or harp.

In order to make the distinction between forced and resonant vibrations (resonant is used here in a rather unspecific sense and is later qualified) consider a child's swing and what happens when we push it. If we put the child on the swing and pull it back and let go we have an example of impulse excited periodic vibration, for the swing will oscillate with reducing amplitude until it stops. If, however, we now give it a push every time it comes to the peak of an oscillation the amplitude will build up, this is a case of resonant vibration and illustrates how much more potent this type of motion is. If, on the other hand, we hold the swing and push it to and fro at a rate other than the *natural* time of swing we shall expend a large force to very little effect and this is an excellent indication of how forced vibrations—unless they result in resonances of other associated parts—are of little harm, except when the forces themselves overload bearings and the like.

Another example, which many readers will have experienced for themselves, is in the case of a car travelling on a corrugated (or washboard) road. The effects of taking such a road at different speeds are illustrated in the artist's impression in plate 1. In the first instance the car is shown as if traversing the road at the critical speed for resonance. Here each rise comes at just the right time to give the car a further push in the direction it is already taking under the influence of its springs. At such speed it is impossible to avoid being aware that the resonance of the mass of the car and occupants on the springs magnifies the vertical motion.

The second illustration shows the car going at a much higher speed (in fact at the anti-resonant condition, though that need not bother the reader at this stage). An infinite speed would also produce the same result. As speed rises from the lowest resonance to the first anti-resonance the motion reduces from intolerable steadily through bearable to a perfect ride.

In describing resonant vibrations there are two terms which are repeatedly used, and which are often misunderstood, yet it is almost impossible to avoid them and these are ' mode ' and ' node '. The word ' mode ' (which in common parlance means fashion or trend) is used to describe specifically the shape or form of motion of an object and, more generally, to refer to those vibration shapes which bear a family likeness. Thus the formal description of the motion of a plucked bowstring would be to call it the flexural mode of the string.

Plate 1. Artist's impression of a car traversing a washboard road; top at the critical speed for resonance; and bottom at the higher speed which is that for anti-resonance.

If a ruler is held on a desk and the free end plucked it also will vibrate in a flexural mode, commonly called a ' flap ' mode—while if it is twisted and released it will vibrate in a torsional mode. As will appear later in the chapter, there are similar but more complex resonant frequencies for each basic type of deflection, so that there are whole families of flexural, torsional, string, bell, longitudinal, etc. modes, all the members of a given family partaking of the same character of motion.

The word ' node ' is applied to any point or line in a vibrating object which is at all times stationary—perhaps it is easier to regard a node as a swivelling or pivot point. For example there are two nodes for a plucked bowstring, one at each end and correspondingly one for the ruler where it is held on the desk. It is equally correct to describe the axle of a seesaw as a node since, though no actual visible part of the seesaw seems always stationary, there is a line along the centre of the axle which does not move at all.

A term of great importance which will recur throughout the book is damping ; again the common usage of this word, even by engineers, is somewhat vague and when damping is not otherwise qualified it means viscous (or Coulomb) damping, i.e. the reduction of a vibration by making the force do work.

A simple practical example of the effect of damping is ‘ ringing ’ a cup or glass to find if it is sound and also of what quality it is. A cracked glass will sound ‘ dead ’ because the rubbing together of the sides of the crack when vibration starts abstracts energy and rapidly stops the motion. A high-quality porcelain cup will sound more ‘ live ’ than an earthenware one because its internal damping is much less. Later on it will be found that engine parts with low damping are all too easily excited by quite small effects to such amplitudes and stresses as will destroy them ; just as a powerful singer can shatter a wingeglass with the right note.

A very convenient term, borrowed from communications engineering, which will be frequently used throughout the book represents, roughly, the lack of damping or ‘ liveliness ’ of a component. This term is *Q* and has long been used by telephone and radio engineers to indicate the goodness of a tuned (i.e. resonant) circuit in respect of amplification and hence selectivity to a given frequency. There is no short equivalent in mechanical engineering terms other than the rather clumsy phrase ‘ sharpness of tuning ’ which is, in any case, not quantitative. Such terms as logarithmic decrement are specific but clumsy, while damping coefficient is no more easily applied and neither gives a clear picture of the effect in respect of resonant vibrations—though useful in calculating forced vibration amplitudes.

The beauty of *Q* is that it is a figure of merit which does not depend on shape or size of the engine part and after a little experience its use becomes second nature in estimating the likelihood of a given excitation producing a lethal effect. Changes in design can be rapidly evaluated if their respective *Q* values are measured.

To take a workaday analogy the aerial of a radio receiver receives signals from many radio transmitters but a tuned circuit, at the input, of high *Q*, selectively resonates to the desired station, to which it has been tuned, so as to produce a signal some 1000 times greater at that frequency than at all the others. Indeed a higher *Q* than this is by no means unattainable.

In an aero engine a *Q* of 2 or 3 means little sensitivity to a specific frequency of vibration, *Q*’s of 20 to 50 are typical of compressor blades and mean a significant increase at resonance while a *Q* of 500—such as might occur in a gear wheel in transverse modes—spells disaster if ever it is energized from even a low vibratory excitation.

This is a suitable point at which to consider more rigorously the differences in the relation between excitation and its effect on the engine. As indicated earlier these mainly arise in five different ways:

1. Forced vibrations.
2. Natural vibrations.
3. Resonant vibrations.

4. Self-excited vibrations.

5. Impulse excited vibrations—which includes ‘ stick–slip ’ excitation.

These are discussed below:

1. *Forced vibration*

This is very easy to understand. It involves the movement of a part by a force so that the part is constrained to move according to its mass in relation to the force. It is rarely significant as a cause of trouble in the engine itself, except when it causes fretting and consequent loss of location of the component, or when bearings are constrained to transmit, say, the force of an unbalanced rotor and suffer thereby.

Figure 1 shows the frequency spectrum, plotted logarithmically in height of response, against a base of frequency of vibration in Hz.

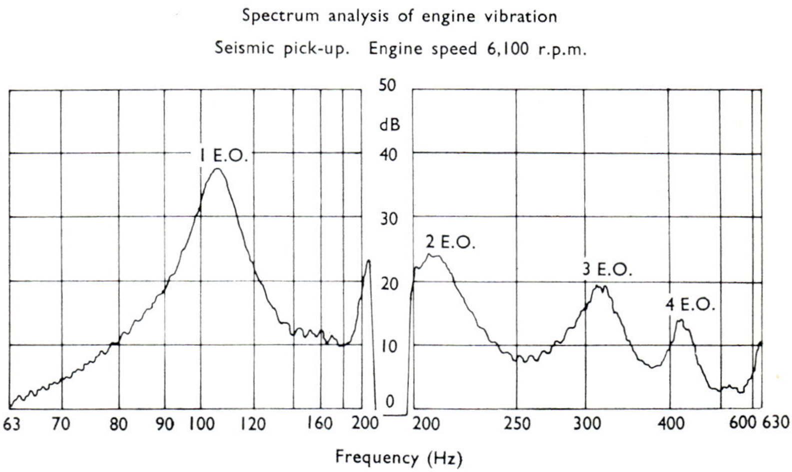

Fig. 1. This shows a typical spectrum plot of an engine vibration spectrum. It shows major peaks at prime shaft speeds (1 EO stands for 1 × shaft speed) and at several of its multiples.

This shows major vibration frequencies at engine shaft speed (1 EO stands for 1 × engine order) and its first three multiples. The plot is made automatically by a Brüel & Kjuer Analyser from a magnetic tape recording of the engine vibration pickup lasting about 30 sec.

Forced movement of a complex accessory may, however, excite resonance within it of individual parts and so overstress them, or produce changes in its normal operational parameters. For example a spring in a fuel metering valve might surge (or resonate) and alter the response of the control. The soundboard of a piano is a practical case of forced vibration (while the strings undergo natural vibration) and is effective because its enormous area couples the small amplitude movement to a lot of air ; in mathematical terms it is an impedance transformer between the strings and the air.

2. *Natural vibrations*

A natural mode of a part is one where distortion, or movement of one bit relative to another, when released gives rise to a cyclic vibration or oscillation which dies away as damping absorbs the original potential energy of the deformation. By definition impulse and stick–slip vibrations are natural vibrations.

While the subject of modes of vibration is dealt with more fully in Chapter 3 this seems a suitable place to deal with harmonics or overtones as they are indiscriminately called. Taking a common example, familiar to most, of a string stretched between two points, its simplest mode of vibration occurs when the middle of the string vibrates with maximum amplitude. This is often called the first bowstring mode. If, however, the centre of the string is lightly held it will vibrate in two halves (producing a note of twice the frequency) and this is called the first harmonic, or overtone, and—perhaps confusingly—second bowstring mode. Similarly it can vibrate in three, four or more equal parts at correspondingly higher frequencies.

In the case of a stretched string these frequencies are very nearly exact integral multiples of the simplest or fundamental frequency. The same holds good with the air in a pipe of parallel section ; but a beam with one or both ends fixed, while capable of producing similar higher modes of vibration does not do this at integral multiples of the fundamental frequency. Hence it is correct to speak of ' harmonics ' of the fundamental in the case of a string or pipe—since they produce harmony with the fundamental (at least in the lower multiples), it is more correct to use the term ' overtone ' which is equally applicable to harmonic or inharmonic higher modes. Throughout this book the lowest of a family of modes will be called ' first ' and not ' zero ' as is used by some writers. This ' fundamental ' is followed by its first overtone and so on. The use of the word overtone is more correct and less ambiguous than harmonic.

It is quite common for a part to vibrate simultaneously in the fundamental and several higher modes of a family or families according to the type of excitation, or it may go singly in one or another overtone. In the analogous case of musical instruments, in fact, it is the rule rather than the exception that many overtones exist and in the piano (for example) the first overtone is almost equal in amplitude to the fundamental while in a bell the fundamental is often barely audible.

Violinists and other string players make frequent use of overtones (they say harmonics) to alter the quality of a note while wood and brass wind players could not cover the whole register of the instrument without them and a bugle depends wholly on overtones for its playing, otherwise it could sound only one note !

3. *Resonant vibrations*

A resonant vibration is one where a part, or an assembly, is excited by a force of such frequency as to approximate to a natural mode of vibration of the part itself—as in the case of the swing pushed at the right intervals. There is an effect arising from the damping in the part which causes the resonant frequency to differ slightly from the natural frequency by an amount which increases with the damping. Fortunately the effect is small, unless the damping is very high, and for practical purposes if the Q is over, say, 200 the effect is almost undetectable and even at a Q of only 4 is barely 1%. This, no doubt, is why most engineers use the term ‘ resonant ’ for both natural and resonant vibrations and indeed the distinction in practice is very small.

A resonant vibration builds up in the part an amplitude which soon reaches a level where the energy input equals the damping losses. For a Q of the order of 45 this may take about 45 cycles, but for a Q of 1200 may require as many as 1200 cycles. Practical examples include the air-column resonances of wind instruments and organs, ordinary speech where the vowels are defined by changing the resonances of the vocal tract, irrespective of the normal pitch, and the ‘ rocking ’ of a car to free it from mud, or in the case of a mob to overturn it !

Figure 2 shows actual straingauge readings of stress in a gear in lb per sq. in. against engine speed. Note the extreme sharpness of the resonances with the three and four diametral modes of transverse vibration of the gear disc. The top diagram shows how, when the sloping lines of frequency of tooth excited force intersect the frequency of the mode in the gear, there is strong resonant amplification.

4. *Self-excited vibrations*

This term is really a bit of a misnomer since literally it implies spontaneous motion. It is, however, endorsed by long usage, and there is no simple alternative. This term means vibrations excited by a force controlled by the motion of the part itself at one of the part's natural frequencies.

A common example is the ‘ singing ’ of a guy wire in a strong wind. Another one used instinctively by most people is the excitation of the cavity modes of the air in a kettle or teapot ; the tapwater running in produces a wide band aperiodic (or random) force which produces a series of notes whose pitch rises as the vessel fills up, so that the proportion filled corresponds to the ratio of the frequencies when empty and partly full. A moment's thought will show that, often without realizing it, most people make regular use of this effect (see fig. 3—Spectra of teapot).

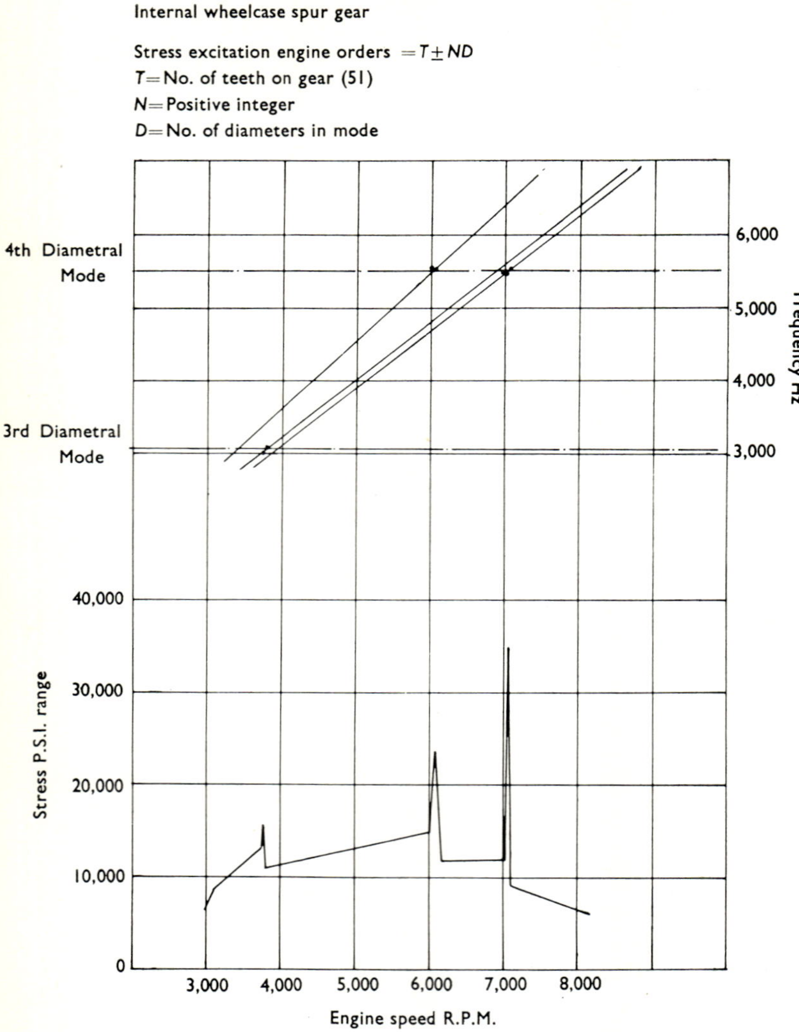

Fig. 2. This plot shows straingauge readings of stress in lb per sq. in. against engine speed for an auxiliary drive gear. The plot at the top indicates that the (very sharp) peaks of stress occur when the frequency of the resonant modes of the gear coincide with tooth excitation frequencies.

In an engine the aerodynamic effects of downstream eddies or change of ' lift ' of a compressor or turbine blade at undesirable angles of incidence can cause this type of excitation, and an everyday example is the fluttering of a venetian blind over an open window.

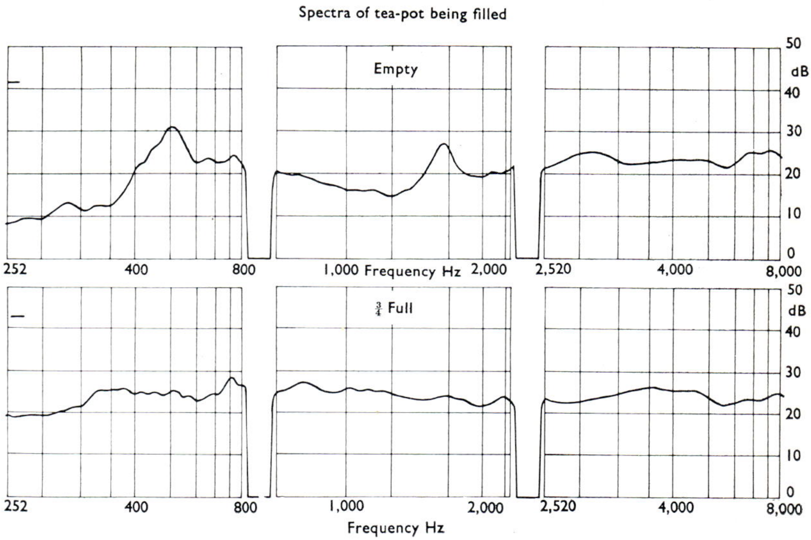

Fig. 3. This spectrum analysis plot of the sound of a teapot being filled shows how the character changes as the teapot fills up. This is because the self-excited cavity resonances alter with the volume of air left. Note how the peak at 500 Hz empty changes to 770 Hz when nearly full.

5. *Impulse and ' stick–slip ' excitation*

This form of vibration arises from a single deflection, a hammer blow or a sudden change of load conditions. Stick–slip vibrations are merely a special case when one member rubbing against another deflects the part until the resistance to deflection exceeds the frictional force when the part springs back and the cycle is continued. The *a priori* requirement is that the static co-efficient exceeds the dynamic—which is nearly always the case.

There are many practical examples, the vibration of a violin string comes from stick–slip motion of the bow on the string, the sound produced by rubbing a wet finger on a glass or a rubber balloon spring to mind at once, as does the tortured shriek of a blunt tool cutting metal in a lathe!

Impulse vibrations in general arise when sudden forces develop as by random rattling of engine parts, or sudden changes of air velocity in a compressor when the engine is violently handled or its control system over-ridden. More normal examples are the whole percussion section of an orchestra—drums, cymbals, etc. or the high pitched ring of the tool in a road drill. Impulse excited vibrations are at natural frequencies.

6. *' Order ' nomenclature*

This seems a suitable place to introduce a nomenclature for the relation of vibrations to their exciting frequencies in engines, which will save a lot of words later on.

When a vibration takes place at the same frequency as the rotation or engagement of an engine part it is convenient to use the word order. Thus if a vibration occurs at the same frequency as the prime speed of the engine we say it is an ' engine order ' frequency, e.g. at 6000 revolutions per minute this means 100 Hz (100 cycles per second). Hence we might identify a resonance as ' pump drive order ' or ' LP ' or ' HP ' order of a twin spool jet engine, or 19 Impeller order of a centrifugal compressor rotor. In the case of an engine's main shaft speed this and its higher multiples are often written 1 EO, 2 EO, etc.

CHAPTER 2

general survey of vibration and measurement

THE types of vibration experienced fall into two major groups (1 and 2 below) and other less usual types and may be stated as:

1. *Forced, general or carcass vibrations*

These occur due to unbalance of rotating parts and imperfection in gears and drives; an excessive amount of this type leads to carcass failure, accessory malfunction and transmission of vibration to crew and passenger accommodation. Most of this is due simply to forced vibration but distress due to local resonances of brackets and controls may also arise.

Such vibration generally covers quite a wide spectrum and includes, of course, that vibration emerging from the engine as sound or re-radiated as such from airframe structural parts, which often act as local resonating amplifiers. This type of vibration is characterized by the inevitable increase with the square of the speed if due to unbalance. If due to transmission imperfection generally, any increase is linear with torque transmitted. Vibrations at frequencies well below that of the engine speed can arise from uneven power generation, combustion irregularities and control system instability. The reduction of these effects requires a higher standard of quality.

2. *Resonant and self-excited vibrations*

These are vibrations of parts of the engine which more generally have effects restricted to the components themselves and lead to excessive wear, malfunction or fatigue failure. This type of vibration is characterized by its occurrence at certain critical engine speeds or conditions. Only rarely does this type cause effects external to the engine except when the part failure cannot be contained. It is a predominant cause of fatigue failure in the compressor, turbine, gearing and shafts of engines.

Figure 4 shows a plot of measured stress near the root of a compressor blade, plotted against the engine speed. The peak at 5200 r.p.m. is due to resonance of the first flexural mode of the blade with an aerodynamic excitation. That at 6750 r.p.m. is due to resonance of the same blade mode with an intake wake at third engine order (3 EO).

Where such vibrations are due to resonance it is often possible to 'tune' (i.e. shift) the resonance to a less used engine speed or so that the frequency of the part is above the excitation frequency attainable at maximum speed. Alternatively it is possible in certain cases to decrease the Q by changes which increase damping.

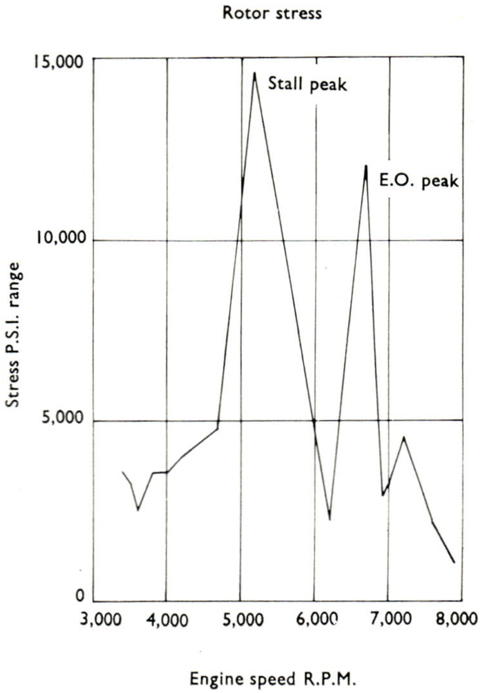

Fig. 4. This shows the stress near the root of a compressor blade plotted against engine speed. The peak at 5200 r.p.m. is due to resonance of the first flexural mode of the blade with rotating stall cells. That at 6750 r.p.m. a resonance of the same blade mode with the wake of the intake asymmetry.

3. *Coupled vibrations*

These occur in such vibration systems as involve sympathetic movement of propeller and engine or engine and external drives to gearboxes and other power offtakes; since these only happen when the whole system involved is present they are difficult to reproduce in testing of the engine alone. These vibrations arise either when resonances in each part of the system coincide or when the whole system creates a new resonant condition. Such coupled vibrations may be flexural or torsional or both.

Since vibrations of this nature often involve interaction between the products of different firms they can create delicate situations with each side rightly pointing out that if the other firm makes a change all will be well! While true, this is usually unproductive, and a co-operative approach is necessary to a successful solution.

4. *Vibrations of gases and liquids*

The former occur in the cavities of the engine, notably in the jet pipes. These more often cause trouble because of the emergent sound, but can cause combustion irregularity and sometimes cracking of sheet metal ducting and parts of the combustion system such as flame tubes and the like.

In the early stages of development of such parts as combustion chambers (flame tubes) and jet pipes it sometimes happens that a gas mode coincides in frequency with a resonance of the chamber. If such coupling occurs then very rapid failure of the component is certain.

More typically—and fortunately more often—forced vibration of the casing by the aperiodic broad band vibration due to the high gas velocity can lead to eventual cracking and failure. However, in the jet pipe, such gas resonances emit a very unpleasant ' howl ' which will probably necessitate suppression in its own right.

In contrast to most types of gas vibration, those occurring in pipes carrying oil or fuel are almost always longitudinal. Most engines tend to be festooned with supply and control pipes—for many gas turbines use hydro-mechanical governing systems, often with engine fuel as the working fluid. The low compressibility of liquids, and consequent low damping, make vibrations in such systems rather severe when they occur, while at the same time, most high pressure pumps tend to be very effective generators of pulsations.

The reader will be familiar with the excellent sound transmission in liquids, say along the waterpipes of his house; some too will have experienced the very severe effects of oscillating conditions in such systems, in the large forces developed (at times enough to burst the pipes) when a tap is suddenly shut off. These are generally colloquially referred to as ' water hammer '—a very apt description indeed.

5. *Damping*

In formal treatises damping is quite simply dealt with since it is generally assumed at a certain value from the outset in working out a problem. In practical work it is difficult to measure directly, but its effect is a major one in most cases. It is no exaggeration to say that if all components were rigidly attached to each other the average design of aero engine would not last many hours before fatigue failure caused it to fall to bits.

There are, however, a few cases, generally due to change in application—or exceptionally to bad design—where a mode of vibration with inherently low damping, is excited and the associated failures are then spectacular! Diametral vibrations of wheels, torsional oscillations of shafts and wheels and radial and slosh modes of gas vibration are the commonest cases of such lightly damped vibrations.

In academic exercises when damping is insufficient to prevent excessive response a dashpot is a simple cure; the equivalent provision

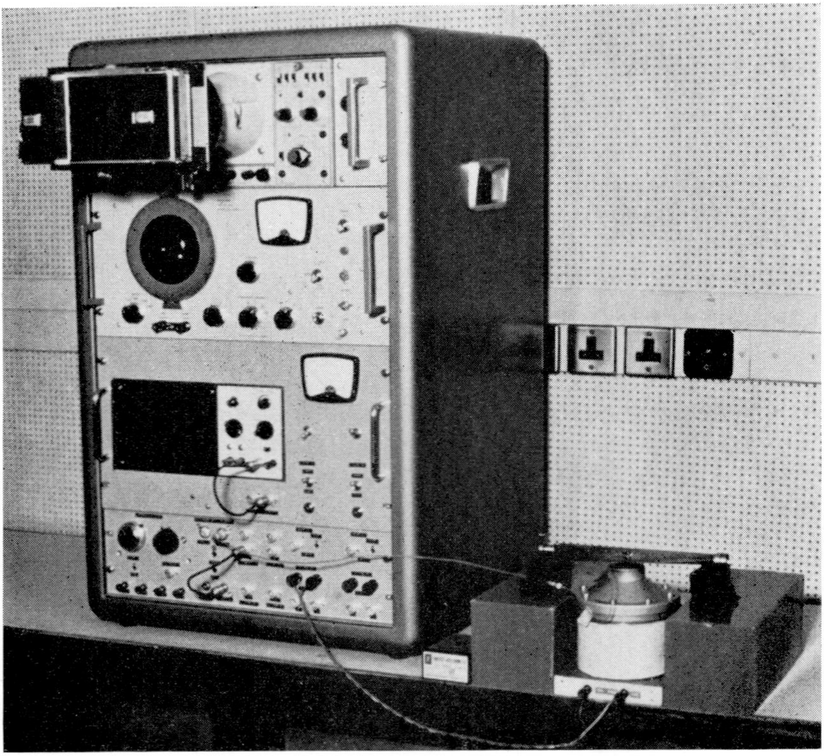

PLATE 2. This illustrates the set-up for checking the Q or damping factor of a hollow steel inlet swirl vane. The horn pressure exciter is set at a resonant frequency of the vane and the power then cut off. The decaying wave is photographed with the Polaroid camera.

in a light high speed engine is likely to cause a complete tear-up of the design. Damping in such cases must cause the vibration to perform work and one is often faced with an effective damper failing from fretting. The classical alternative of a harmonic damper (spring supported countermass) is rarely of practical use in aero engines, with their wide speed range, since it can only suppress resonance at one frequency, i.e. at one engine speed, while it allows resonance at two other speeds.

There are under development alloys of copper and manganese or of nickel and manganese which can be produced with very large internal damping. Their other physical properties are not yet suitable for many applications. Such special alloys apart, internal damping in metallic components is almost always insignificant in amount.

Plate 2 shows a set-up for determining the damping, by measuring the Q, of a hollow compressor inlet guide vane. The vane is excited at an appropriate resonance by the horn unit and the decaying wave resulting when power is interrupted enables a calculation of the Q (or the logarithmic decrement) to be made.

Measurement is now discussed which is divided in the same manner as the previous sections on vibration.

6. *Forced general or carcass vibration measurement*

One of the greatest difficulties in measuring these vibrations is the absence of any fixed reference point and the enormous variation in different parts of the carcass. No attempt is normally made to measure other than the movement of the carcass in space—some of which movement may be due to causes outside the engine. For reasons dealt with more fully later it is usual to use seismic pickups for monitoring general engine vibration and accelerometers for local measurements related to trouble in specific places. It is desirable to site the monitoring positions for the former where these receive a preponderance of engine order unbalance, i.e. on some place such as a bearing housing casting or other more or less solid part, since mounting of the pickup on sheet metal fabrications introduces colouration of the engine frequency by many higher ones due to local resonances.

Only when wear or cracking, arising from the local resonance, is being investigated does one explore in detail the vibrations at such places. However, these must be measured with great care since it is relative movement which is usually of interest and an attempt to sort out the relative movement of two parts a few inches apart by subtracting (vectorially) the outputs of two pickups which show differing spectra and extending this exercise to two or three planes is fraught with difficulty—if not foredoomed to failure. Most vibrations of this nature which lead to local fatigue are investigated with straingauges; those leading to wear by local stethoscope techniques or stroboscopic examination on the running engine.

7. *Resonant and self-excited vibration measurements*

Since such vibrations are usually characteristic of specific components their measurement is generally done by the use of straingauges and proximity probes, chiefly the former. The difficulties here are associated mainly with leading the signals out of the engine, particularly from rotating parts. Environmental conditions for straingauges and pickups are severe—if not hazardous—and an accurate knowledge of the ambient conditions beforehand would save many abortive tests. There is likely to be an increasing use of optical and radio methods for observing the vibrations of parts not easily accessible as time goes on and the advent of fibre optics and Gunn diodes makes itself felt.

8. *Coupled vibration measurement*

Coupled vibrations which have resulted in failure of a shaft or casing are preferably investigated by straingauging the defecting

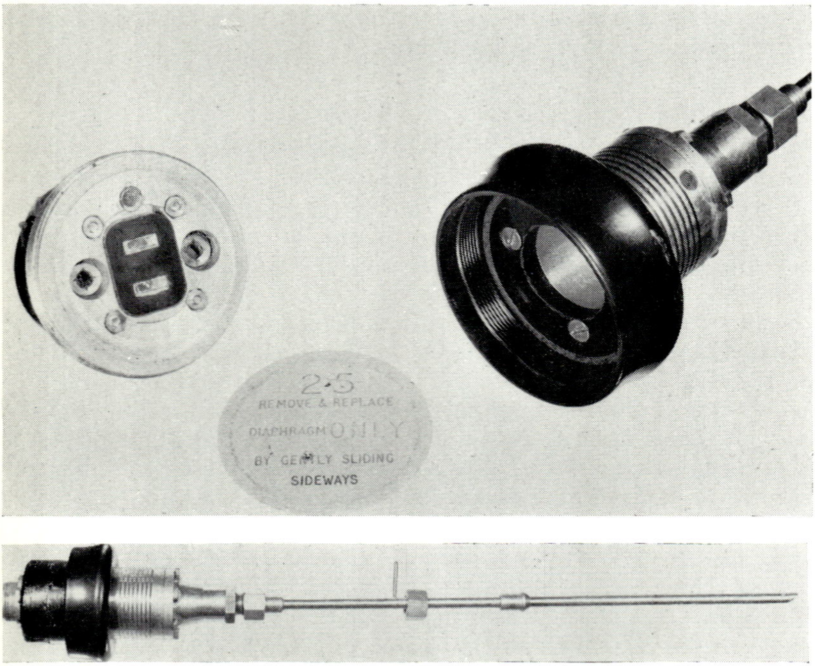

PLATE 3. This shows a simple electrical stethoscope which may be used to listen to what is happening in an engine compressor. The pressure inside is held back by a thin diaphragm which transmits the sound to the earpiece, the output from which is amplified electrically.

component and identifying the frequency. However, in many cases this is inaccessible for such measurements and recourse must be made to vibration pickups for flexural modes and torsional accelerometers for torsional modes. The latter instruments might also be required in order to test any suggested cures for the failure when its frequency has been determined.

9. *Fluid pulsation measurement*

Such simple methods, as are used in the school physics laboratory, for showing gas movement by the use of lycopodium powder, or tufts of cotton wool on threads, are clearly useless in a running engine. Hence it is usual to make use of pressure tappings led to suitable transducers, or anemometers, in order to determine the frequencies and spatial distribution of the gas movement.

Plate 3 shows a simple stethoscope, which has been used to listen to what is happening in an axial compressor. The sounds inside the compressor are conducted up the tube which can be situated between two stator blades. The internal pressure is held back by a thin metal diaphragm, through which the sound is conducted, to the telephone earpiece. The signal from this is amplified, and perhaps filtered electrically, being then heard from headphones or a loudspeaker. It is a much more pleasant and more efficient way of finding out what is going on inside than putting one's ear to the casing.

Pulsations in pipes carrying liquid are generally investigated using pressure transducers; paying particular attention that the fitting of these does not significantly change the length or volume of the system. Sometimes, however, it is possible to make rather a neat use of strain-gauges for the purpose and avoid making any contact with the liquid at all! If suitable straingauges are fixed to the pipe, preferably circumferentially, the pipe will become a pressure transducer without any dead volume. Now that small foil gauges are available this is practicable even on pipes of one quarter of an inch diameter. While not very suitable for quantitative measurements, because of end restraint conditions, it is possible by this means to find the frequency of the pulsations and the engine conditions at which they occur.

CHAPTER 3

typical modes of resonant and self-excited vibrations

THIS chapter deals with the main types of vibration patterns met with in aero engines.

1. *Cantilever vibrations in blades*

The majority of aero engine compressors and turbines have their rotor blades held in the supporting discs, or drums, by pin joints, dovetail or firtree features so that at low engine speeds they are not fully encastered, i.e. rigidly held. This is not only because replacement of damaged blades would otherwise be difficult; but also makes it possible to reduce the vibratory stresses at lower engine speeds—where the matching is less satisfactory and aerodynamic excitation might otherwise impose unacceptable stresses. For this reason the frequencies of the various cantilever modes of the aerofoil section change with engine speed in addition to that alteration due to centrifugal force on the aerofoil.

A freely pin-located blade, with a designed clearance between pin and blade, has only one mode possible at zero speed and this only if it is at the bottom of the rotor. This mode is the pendulum one and may typically be of the order of 30 Hz. This mode can occur also in the running engine at cranking (or starting) speeds but is rarely excited, though instances have been met where this has caused ' ratchetting ' of the blade pins and consequent pickup on the mating surfaces. The commonest encastered mode is that of first transverse resonance where motion is normal to the chord of the aerofoil, this is often called first flap for obvious reasons (see plate 4). The only node location in this mode is along the root of the aerofoil. This mode is the easiest to excite and it is desirable that the frequency, in the engine, should be above twice engine rotation frequency at all running speeds from idling upwards. Because a pin-fixed blade can vibrate with the pin rolling at one frequency, and should the pin ' freeze ' at a different frequency, the former is sometimes called ' first pin rolling ' and the latter ' first flap '.

Higher harmonics in the transverse modes exist at second, third, fourth flap, and so on which exhibit a corresponding number of nodes along the blade as shown in fig. 5 which shows typical nodal lines for a compressor blade in the flap family. It is understood that on either side of any nodal line the instantaneous movement direction is reversed;

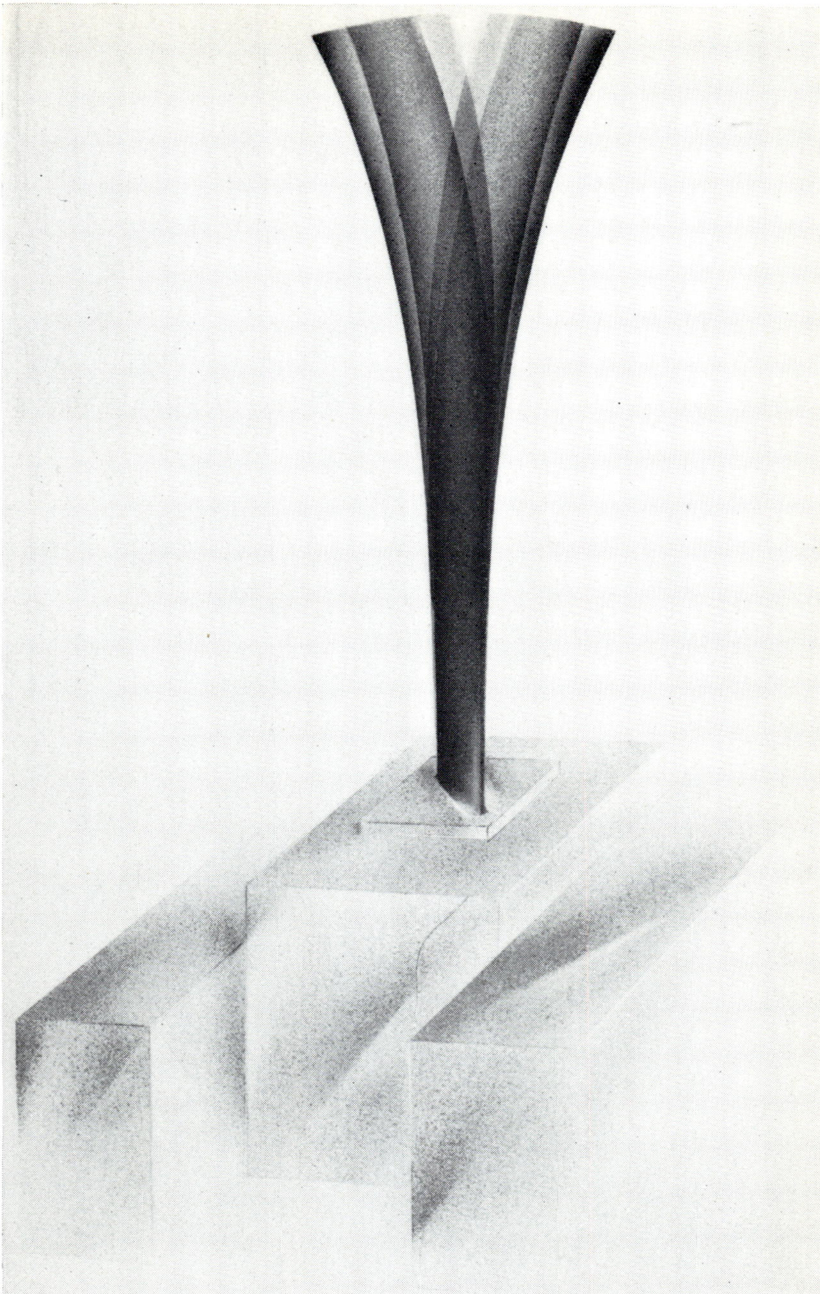

PLATE 4. An artist's view of the movement of a compressor blade in first flexural (first flap) mode.

PLATE 5. Artist's drawing of the motion of a blade in the second flexural (second flap) mode. Note that the outer node is a swivelling point—not a point of high bending stress.

the other principal cause of blade resonance in flap modes is coincidence of blade frequency with rotating cell stall patterns. The higher harmonics are sometimes excited by passing order of the wake of previous stators but this is less likely with typical designs.

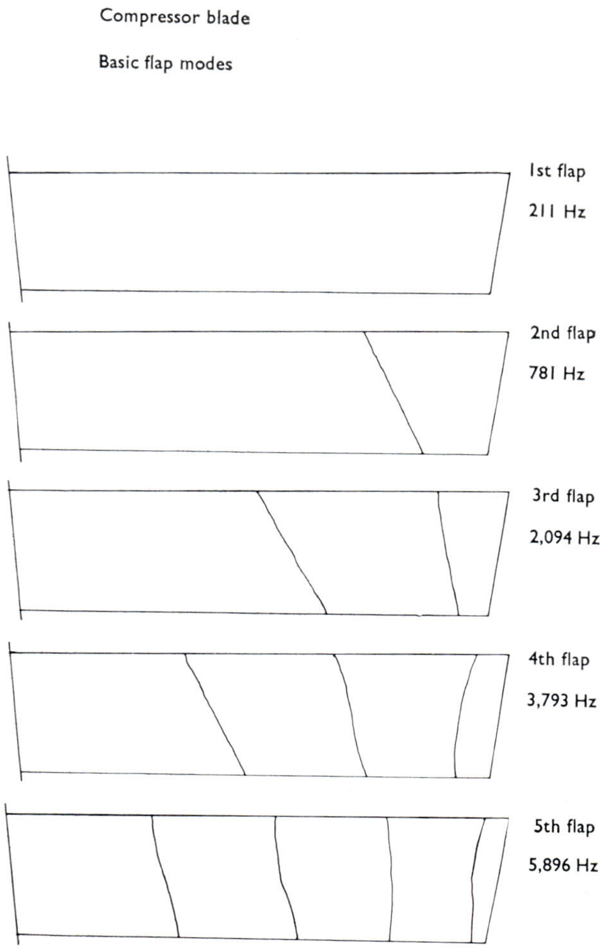

Fig. 5. Drawing of the nodal lines of the basic flap modes of a compressor blade, as revealed by sand patterns.

Another series of blade modes is that of torsion. See fig. 6, which shows typical torsional nodal patterns of a compressor blade which in this case is tapered and twisted. Here again the nodal lines are distorted from those of a uniform cantilever and to a greater extent than are flap modes. It is more usual to find that torsional modes are self-excited by abnormal angles of incidence than by resonance with intake wakes or rotating stall effects. The damping of torsional modes at lower engine speeds may well be considerable, but in the higher part

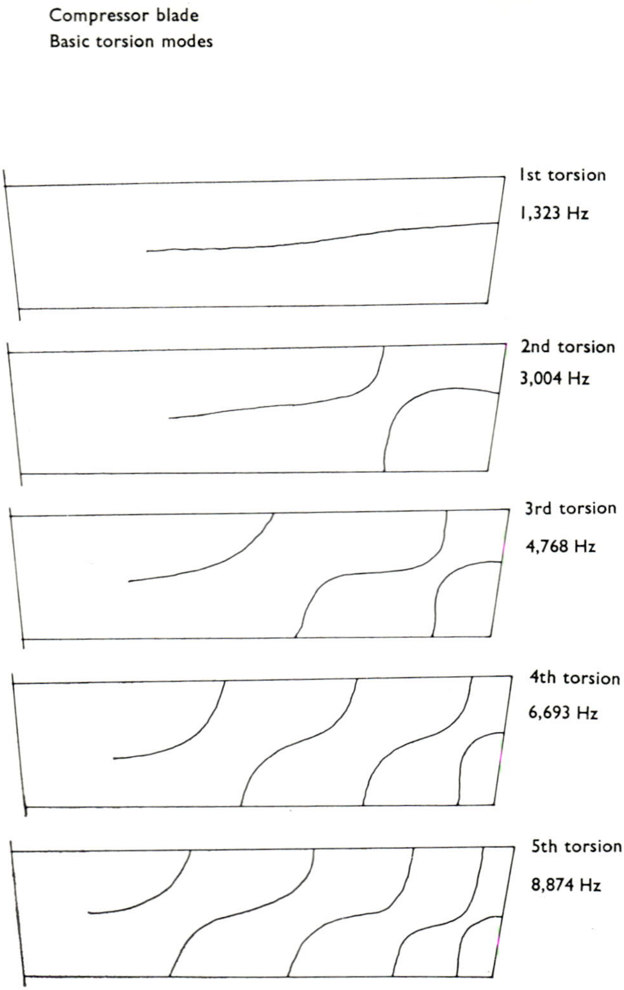

Fig. 6. Drawings of the basic torsional nodes of a compressor blade as revealed by sand patterns.

of the speed range they benefit less from aerodynamic damping than flap modes—and scarcely at all if the vibration is self-excited by the airflow.

In cases where resonant excitation is the cause of trouble it is often possible to tune the blades out of resonance by cropping the tip of the blade, see fig. 7 which shows the effect, on a complex ‘ lyre ’ mode, of changes to blade geometry. This type of resonance can cause tip splitting, which is often prevented by cutting back the tip of the blade; or by changing the pin clearance or fit of the root dovetail or firtree.

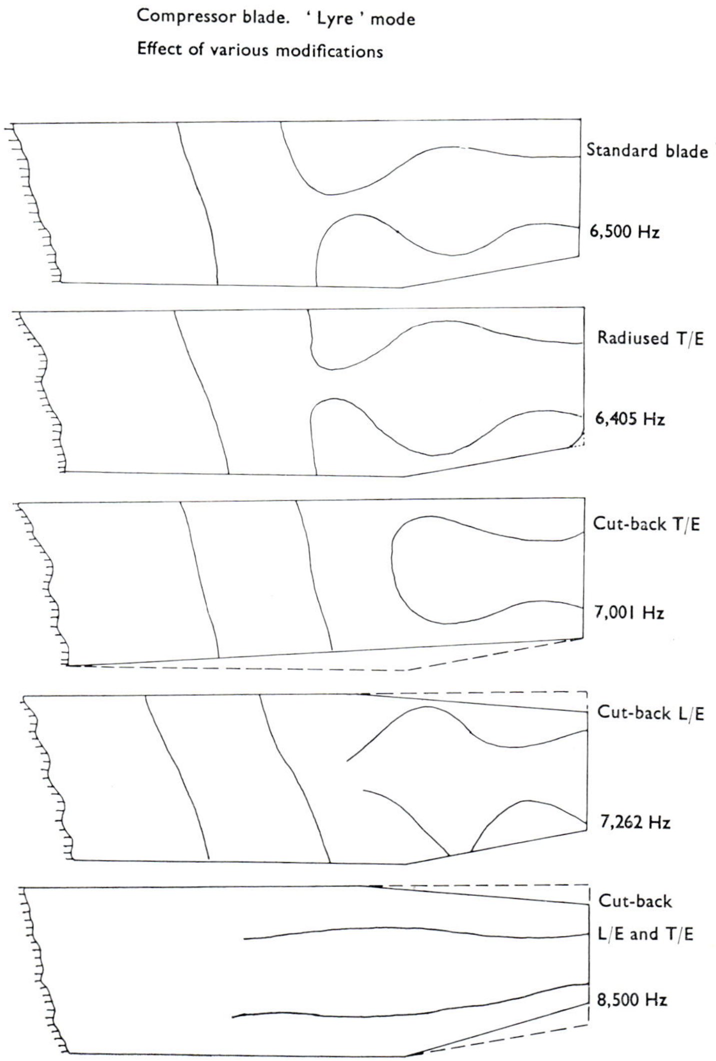

Fig. 7. Drawing of the nodal patterns of one complex mode of a compressor blade, showing the changes made by modifications to the blade.

However, this cannot be pursued too far, in the case of rotating stall excitation, since if applied to certain stages the stall may change its frequency to a different one suited to resonate to a controlling stage's own frequency and you are back where you started—or worse off still.

While the only way of measuring the actual running resonant frequency of a blade is to straingauge it, or calculate the frequency by one of the matrix methods—a great deal of supplementary work can be done more rapidly and cheaply by encastering the root, in a manner similar to that in the engine, either by holding it in suitable jaws in a vice free from damping or by casting in zinc alloy. These methods are used in determining the order of changes which can be wrought by dimensional alteration, confirmed afterwards by engine tests.

The third series of modes possible with blades is the edgewise one, i.e. motion parallel to the chord. In compressors this is generally so high as to be rarely encountered but in turbine blades this is not so. Most turbine blades carry a shroud, with seals, at the tip in the interests of low overstage leakage and the mass of this brings edgewise modes lower than would be the case for a compressor blade. In addition upstream of the turbine entry there are several wake generators of relatively moderate frequency, e.g. discharge nozzles, nozzle guide vanes and casing to bearing support struts while the gas velocity is necessarily high.

2. *Bowstring vibrations in blades*

Single unshrouded compressor blades cannot develop a bowstring mode, but stator blades are often shrouded and the individual blade shrouds joined in packets, or the shrouds may interfere anyway. Turbine rotor blades are generally shrouded and sometimes packeted by welding in two's or three's to increase the flap frequency. Turbine stator blades, or guide vanes, are commonly held at both ends. The causes of vibration excitation are the same as stated in § 1. See plate 6 an artist's impression of motion of a packet of blades in bowstring. The difference shown in amplitude is characteristic of such a packet where the individual blades differ slightly. Bowstring frequencies are higher than cantilever ones for the same aerofoil and the stress concentration at the encastre points are generally several times higher; for some reason stator blades show a greater propensity for self-excitation effects than rotor blades. Because of the absence of centrifugal effects the frequency remains almost constant throughout the whole speed range of the engine. Rotating stall and, of course, surge seem especially prone to cause high bowstring stress in compressor stators. It is unusual to find the higher harmonics of the fundamental bowstring frequency present, although severe aerodynamic disturbance sometimes brings in the second bowstring mode. Since the blades are near to double encastered beams the successive higher modes are not integral multiples of the fundamental.

Besides rotating cell stall, surge or near surge conditions cause very high stresses in compressor blades and because surge is an unstable phenomenon it is often difficult to get repeatable results when attempting instrumented measurements, which implies a lot of repeat testing for valid conclusions.

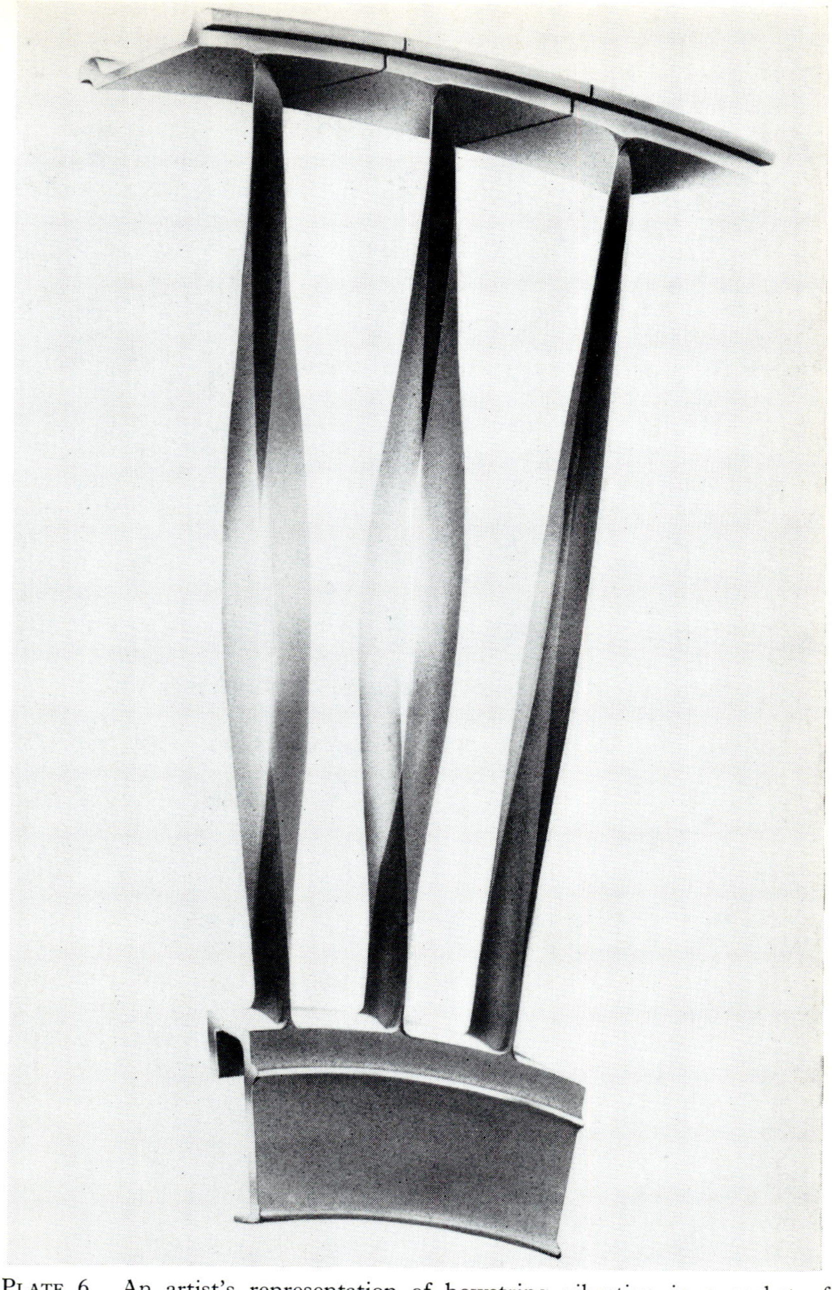

PLATE 6. An artist's representation of bowstring vibration in a packet of compressor stator blades. The variable amplitude shown is no accident but does occur because individual blades have different resonant frequencies.

As well as causing high stress, under unfavourable conditions, bowstring modes of vibration are particularly prone to cause excessive fretting at the attachments of stator blades and nozzle guide vanes because of the multiaxial movement which usually results in practical cases, since it is difficult to design zero clearance or interference fits and still build the engine. Here the designer is in a cleft stick since a very tight fit results in low fretting but correspondingly high bowstring resonant stresses, while a slack fit which damps the assembly nicely, enhances fretting which then increases the clearance, and makes the fretting still more rapid.

As an example of the above dilemma the writer was involved in a case where the same engine and blades required opposite treatment in Civil and Military versions. The problem in the Civil (and Industrial) engines was that of excessive stator blade platform fretting leading to loss of location at lives from 2000 to 20 000 hours. Primary failures of these blades from fatigue was not likely. In this case the inner end shrouds were encastered firmly to a metal ring which completely suppressed flap modes. The self-excited (wind) vibration which then remained was not harmful.

The application of a similar ‘ cure ’ for the military engine trouble, which was primarily a failure of the blades in fatigue, was unsuccessful. This arose because the self-excited (wind) vibration now caused severe bowstring excitation at certain flight conditions not met with by the civil or industrial engines. Though the movements in bowstring were smaller than they would have been in flap the stress concentration in the platform went up and the blades were nearer to failure than when they were in separate packets before. The short life of the military engine meant that fretting was not such a serious problem as blade fatigue.

3. *Bowstring vibrations in pipes and wires*

Because of the type of design, an aero engine has a large number of pipes interconnecting air and hydraulic units, and the majority of these pipes have of necessity to be multiplanar. A consequence of this is that the vibrations of these pipes are not only less simple to understand, but are a great deal more easily excited than they would be were the pipes straight. It is often easier for the engineer to comprehend the manner in which such pipes vibrate if he realizes that a movement of one part of the pipe in one plane produces not only a deflection in that plane but a couple which will produce other deflections in different planes from the first. For this reason it is impracticable for the designer to calculate the resonance frequencies of such a pipe if it has any degree of complexity, and sound room tests on the pipe, assembled as it will be on the engine, are very important. Unless the pipe is under heavy tension and is straight the higher modes are not integral multiples of the fundamental.

Because of the fact that many such pipes are liable to suffer considerable standing stresses—both from thermal effects of unequal expansions and from errors in the geometry of pipe and engine—the occurrence of relatively low alternating stresses due to vibration are often very serious. It goes without saying that the failure by fatigue cracking of a pipe carrying high pressure fuel is almost certain to cause a major fire within the engine or the engine bay, and may well result in the loss of an aircraft.

It is unfortunate that the designer, guided often by an instinctive desire for symmetry, tends to place his restraining clips on longer pipes at such points as are a simple division of the length. Unfortunately this is the worst possible thing he can do, since in so doing, while he will inhibit the first bowstring mode of the pipe, he will leave entirely unrestrained that of the second (if his clip is in the middle) and correspondingly all higher modes according to the division of the pipe length by the clips. While less tidy it is better to put pipe clips at non-integral divisions of the pipe length, and to take great care that they are effective in more than one plane. Consideration should also be given to the conditions that will occur should a clip fail between engine servicing periods, which would detect the fracture of the clip. It is vital that a pipe that failed the clip should not be likely to suffer in any rapid manner before discovery of failure, e.g. during the remainder of a flight.

In regard to the excitation of such resonances in pipes, it is important to remember that, while the main excitation is engine unbalance and its lower harmonics, there are frequently other causes of excitation, at much higher frequencies, which can occur in the running engine. A particularly likely cause of vibration in other than a straight pipe, is the effect occurring from pressure pulsations in the fluid being conveyed. Since the engine fuel is normally pumped by a swash plate, or occasionally a gear pump, it follows that there will be considerable fluctuation at plunger or gear tooth order within the pipe. In such cases, when the pump might cavitate (typically if a booster pump in the aircraft fuel tanks fails) much higher excitations and frequencies may occur.

There is also another potent cause of pipe excitation, and this can be quite spectacular in its effects at times ; it occurs sometimes that a pipe supplying lubrication to a gear train is so placed that one end of the pipe is moving at gear tooth frequency, or one of its multiples. In such cases a very powerful excitation of the pipe occurs at frequencies which are high enough to cause failures of even quite short stiff pipes. Cases have occurred where a pipe, no more than 6 in. long, lubricating the main reduction gear of a propeller engine has failed in less than 50 hours from such a cause, and the designer would not normally consider subsidiary clipping of a pipe of this length. However, a slavish adherence to multiple clipping on such grounds, whether high frequency excitation is likely or not, can do more harm than good, and the author is only drawing attention to the fact that it is important to consider the likely sources of excitation in any given case.

The flexible wires which connect most of the electrical accessories are not, in general, likely to suffer much from vibration as long as they are effectively anchored on the units to which they run, so that forced vibration of the wire does not break it down at the joint. However, the firewire warning system of the engine consists of a number of runs of metal-sheathed mineral-insulated wire, which can easily suffer from bowstring vibration leading to false fire warnings. Calculation—and indeed measurement—of such effects are difficult, because while a relatively rigid pipe is not much changed in frequency by any likely tension applied by the engine, the same is not true of firewires. The relatively small section of these wires means that they can be assembled to the engine in almost any state of tautness at the whim of the particular fitter, and hence can have almost any frequency, which will not be predictable by the designer. An expensive way of dealing with this problem would be to wrap the whole length with a serving of soft wire which is very effective in damping the vibration, but this is usually impracticable. The usual solution is to clip at frequent intervals, and to make sure that where the wire changes direction, it is not restrained from accommodating the local tensions. With a taut wire the higher modes are simple integral multiples of the fundamental.

4. *Disc vibrations in compressor and turbine wheels*

The reader acquainted with classical dynamics will be familiar with the Stodola type of excitation, and those not so acquainted will find ample information in the references of the Bibliography. Put briefly, such type of excitation occurs when a rotating disc spins past a fixed force or deflection at such a speed that one diametral node of the disc arrives regularly at the force point in the right phase relation to perpetuate the vibration. In such cases the disc (and any blades attached to it) can resonate to an amplitude limited only by the damping, and the effect will persist as long as the shaft continues at the same speed. The higher diametral modes are not integrally related to the fundamental two diameter mode.

Since a single force occurring once in each engine revolution can be considered as a Fourier series, it follows that as the engine runs up the speed range, there is a speed for each diametral mode of the disc, which may excite resonance. For this reason it is quite impossible to design a practical disc which will not be to some extent excited into diametral resonances in this manner within the running range of the engine. However, the higher harmonics of a single engine order force are generally weak and the effects held safely in check by the damping. This is particularly the case when more than one stage of turbine wheel is used since there is usually appreciable damping at the interface between adjacent turbine wheels.

Because it is particularly difficult to visualize the motion of a disc in diametral vibration, unless it can be observed with a stroboscope at high amplitude, which many readers will not be able to do, an artist's impression of the motion appears on plate 7. It should be understood that, since the vibration must proceed round the disc so to finish

Plate 7. An artist's impression of the three-diameter mode of a disc. Note that the maximum skin stress, due to bending, occurs at the anti-nodes (where the curvature is greatest) and not at the nodes where only swivelling takes place.

up in the same direction of flexure, whichever route round is taken, it is only possible to have an integral number of diameters. An odd number of radial nodes would leave one in the absurd position of having one part of the disc going in two opposite directions at once!

Yet the author has heard experienced engineers say that, for example, an impeller with 19 vanes can resonate in a 19 radial line mode ; if the test is performed it will be found that the mode which has all the vanes apparently nodal has a diffuse pattern over a part of the disc, and an accelerometer will show that there is one part where there is 180° phase change over two vanes, not one as with a true diametral.

Note that the maximum skin stress, due to bending, occurs at the anti-nodes (where the curvature is greatest) and not at the nodes, where the surface merely swivels.

It is usual therefore to endeavour to design the wheels so that their lower modes, say up to about six diameters, are not excited by forces between first and third engine order within the part of the running range when significant power is being developed, i.e. from about 60% maximum speed upwards. Even this is often rather a difficult condition to secure without excessive thickness and hence weight. Typical causes of excitation are a blockage in one of the passages upstream of the turbine, caused perhaps by some part of the canware having become detached and lodging against the nozzle guide vanes ; or the failure of a burner; less frequently the down stream wakes of asymmetries round the periphery of the nozzle guide vane area of the engine, and but rarely by the nozzle guide vane frequency itself, since this is usually too high.

It is fairly easy to use computer programmes to calculate the frequencies of the diametral modes of such discs, allowing for the effect of the blades and the centrifugal effects inevitable, but it is very difficult to estimate the degree of excitation in any given engine. It is nearly always a requirement that straingauging be carried out in order to determine the degree of excitation that a likely blockage or combustion failure might cause. Failures from such types of vibration are, fortunately, not generally rapidly propagated, and frequently start as cracks in the bucket grooves or firtree blade root features, and are found on overhaul ; although it has happened that a multiple blockage, say a triple blockage at 120°, has caused a rapid cracking and bursting of the disc, these instances are extremely rare.

In theory, at least, it is possible to excite umbrella and ring modes in turbine and other discs. In the sound room these are very easy to produce, although in the running engine the lowest ring mode, i.e. the umbrella type of motion, is heavily damped, since it involves a heavy reaction on the hub of the disc which will react through the bearings and structure to provide damping. There is another possibility, and that is with a double umbrella mode of two stages coupled to each other at the hub, and probably part way along the diaphragm; which has not got such a reaction at the hub—this being the one where an umbrella mode of the two discs occurs with an anti-phased relationship between them. This can be demonstrated in the sound room if the right frequency coincidence is present between the two discs, but has never been observed in the author's experience to occur in an engine. One of the fortunate safeguards against the occurrence of such types of disc vibration is the fact that they require for their production an in-phase force round the whole circumference of the disc which is itself pulsating at the proper frequency. This would typically imply axial gas pulsations, e.g. from burner pressure fluctuations, or central shaft axial vibration at frequencies well above 100 Hz.

Generally, but not invariably, the diametral modes have a pattern that is stationary in space (since the force field itself is stationary) and all points on the disc scan this field in turn.

Plate 8 shows sand patterns of specific examples of the first three families of vibrations of a uniform disc.

Neg. 1

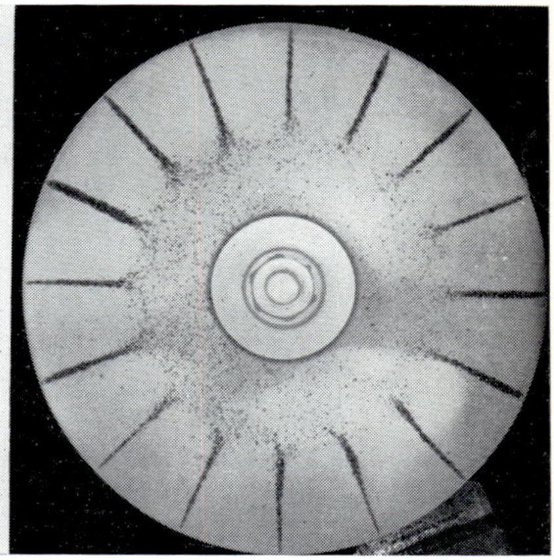

Neg. 2.

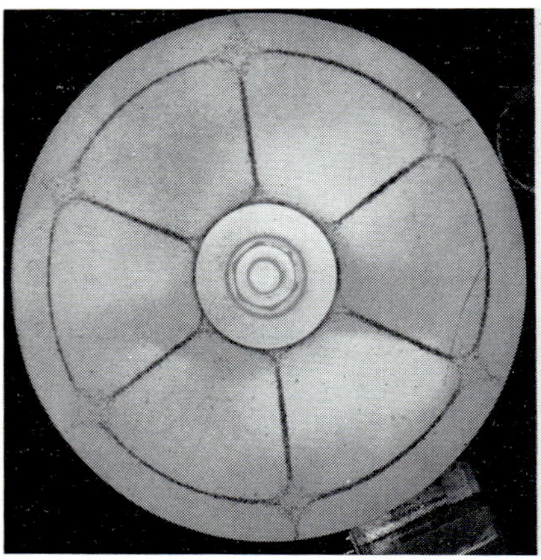

Neg. 3.

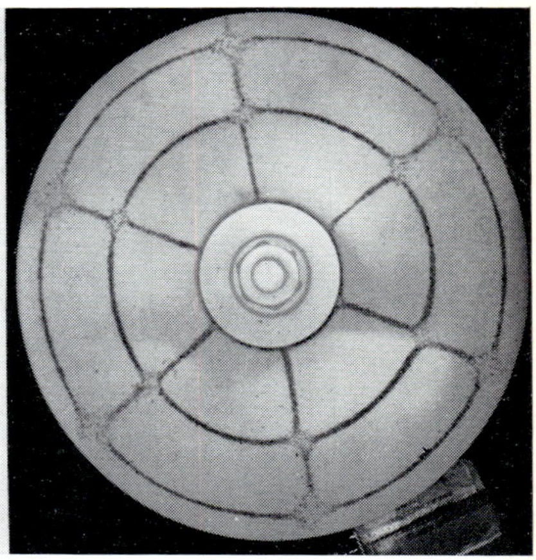

Neg. 4.

PLATE 8. Sand patterns of nodes on a steel disc. Two examples of the first series, six diameter and eight diameter are shown. One example of the second family with three diameters and one ring, and the third family with three diameters and two rings.

It should be noted that there are similar families of diametral modes in compressor or turbine disc with blades. The frequencies of the lowest diametral family start above the umbrella (or first ring) frequency. The second family of diametrals with a central ring and one outer ring may, or may not, overlap the first in frequency, depending on the shape of the disc and the relative contribution of the blades. Cases are known where the first family is asymptotic to first flap of the blades; the second series to first torsion and the third to first edgewise, but it is equally possible that this does not happen.

Within the writer's knowledge it is unusual to find significant excitation above the second family and this latter only when the blades are massive in relation to disc stiffness, e.g. heavily shrouded turbine blades. In some such cases the first edgewise frequency of the blade may be below its first torsion.

There are, of course, other modes of vibration of discs which occur when the disc is held at the edge and not at the centre. In the writer's experience, present aero engine practise has not shown indication of such modes, since it is more usual to take out the torque from the disc at the hub. Such considerations do not apply to drum construction, with which the writer has no experience. The disc spacer ring design common in axial aero compressors may predispose to the second ring family, the outer ring position being fixed by the outer spacer and in these cases the first family may not occur at all.

5. *Disc vibrations in impellers*

In the case of centrifugal compressors the single, or double sided vaned disc is capable of all the modes of any other disc. However, the stiffness of the vanes usually inhibits ring modes from practical consideration. Diametral modes do occur, particularly in single-sided impellers and for this reason it is usual to have an odd and preferably a prime number of vanes. Since there are inevitably diffuser vanes stationary near the tips of the vanes, a very heavy excitation exists at diffuser vane and multiple orders. Unfortunately the damping in an integral vaned impeller is considerably less than that in an axial compressor disc and blades and diametral and vane modes occur very easily.

Since it is not possible to avoid these excitations a careful check needs to be made that the likely resonances do not occur, say, with first or second diffuser vane order within the high power part of the speed range, which fortunately is narrower than in the case of axial compressors. Diametrals of the backplate up to 10 diameters are readily excited and are not very heavily damped, nor are most of the lower vane modes.

In many successful designs there are steel rotating guide vanes at the eye of the impeller which are designed to spring on to the edges of the impeller vanes, and fretting at the interface reduces the Q of the assembly—until the interference disappears after long running. Sound room tests for frequencies of resonance must be carried out in this case both with an without such interference. It is fortunate that the need to provide seal ribs on the backplate to restrict leakage there help greatly to stiffen the disc and raise the backplate diametral modes to safer frequencies.

6. *Disc vibrations in gears*

A perfect pair of gears rolling on each other with the shafts parallel can produce no forces normal to the plane of the gear disc; there would therefore be no excitation for diametral vibrations. One might also expect that because of the tooth load there would be a high degree of damping of such an excitation by friction between the teeth. Practice has shown, however, that with full hydrodynamic lubrication there is a remarkable freedom to vibrate with small amplitudes at high frequencies.

The forces which arise with a practical (imperfect) pair of gears on axes not truly parallel (and what aero engine has rigid casings?) can be significant. There is rarely any trouble with swash or eccentricity of either gear since such frequencies are relatively low. At the same time imperfect tooth form or improper tip relief for the torque at a particular condition combined with lack of parallelism of the shafts can produce significant loads normal to the plane of the disc.

Johnson has shown how to calculate the frequency of excitation from the tooth frequency, while the high Q possible in a typical engine is shown by a test made with straingauges on an auxiliary drive gear of an engine when loaded by the auxiliary—see fig. 8 which shows the straingauge output, converted to stress equivalent, against engine speed. The high Q and the consequent sharp tuning are well illustrated by this actual test, and it is clear that such resonances can be quickly lethal, although the chance of running any length of time at the critical speed is correspondingly less. The diametrals in such cases can either be fixed in space or rotating with the disc, the latter being more often found.

Since gear trains with variable load are general, and tip relief cannot be correct at more than one load, while tolerances on components and deflection of engine carcasses lead to shaft malalignment, it is always safer to avoid flat parallel gear webs whenever possible. To starve the teeth of lubrication would greatly increase the damping but is not usually favoured on the score of life!

7. *Tube and cone modes in casings and nozzles*

Such tubes, or pipes, as are part of the engine casing; jet pipes, flame tubes, final nozzles and the like are generally so stiff in simple bending as to inhibit the flexural modes as a beam. However, they are so thin, on account of weight limitation, that they can be quite easily excited in modes which are variously described as ' bell ' modes, dustbin modes, etc.—these have axial nodes at the ends of diameters.

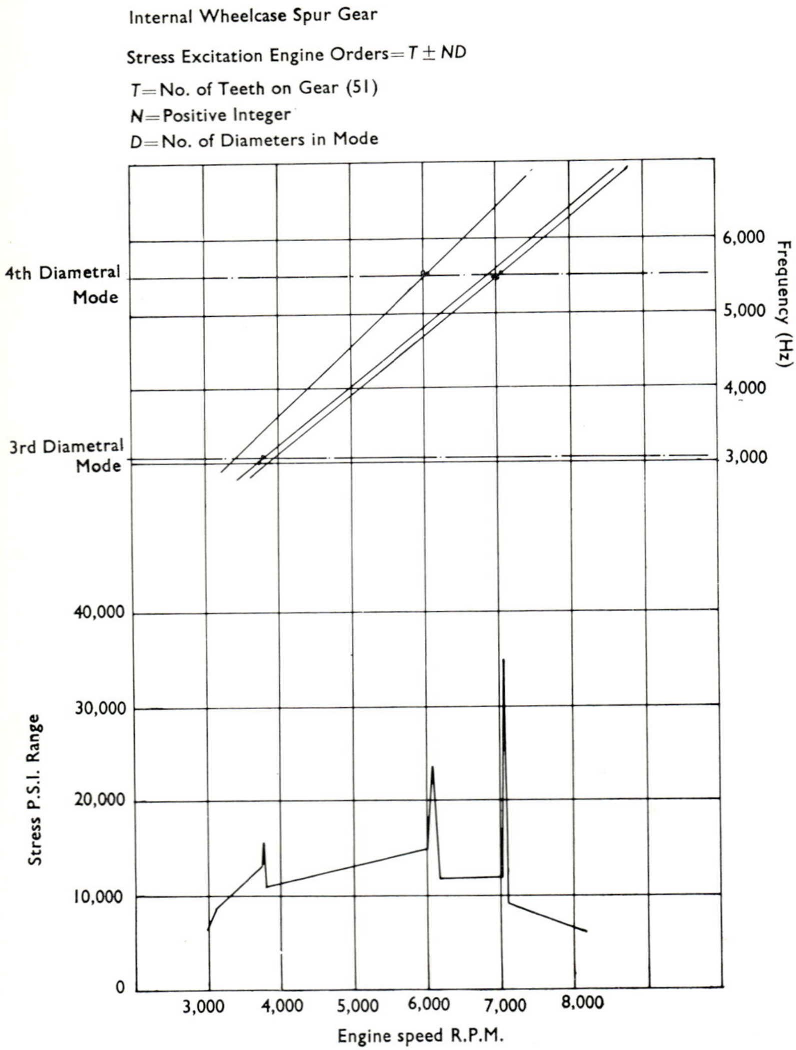

Fig. 8. This plot shows straingauge readings of stress in lb per sq. in. against engine speed for an auxiliary drive gear. The plot at the top indicates that the (very sharp) peaks of resonant stress occur when the frequency of the resonant modes of the gear coincide with tooth excitation frequencies.

There are higher families of such modes with nodal rings at simple divisions of the length. The latter being common on tubes but uncommon on cones of large angle.

Such modes can be excited by engine unbalance and such excitation may be enhanced by local masses of accessories mounted on the pipes. This is the more powerful if the designer has chosen a symmetrical angular disposition of more than one accessory. These modes rarely cause any general distress to such components but may cause cracking where the vibration is locally restrained, at a joint flange or mounting pad or similar change of stiffness.

Another powerful cause of excitation of such types of vibration is excitation from gas vibration (see § 9) and a third cause, notable chiefly in the case of flame tubes, is excitation at natural modes of the part by the ' white noise ', i.e. random noise, produced by the high velocity of the gas and the combustion process. In the case of the latter effects no amelioration by tuning is, of course, possible and increasing the thickness or stiffness is the only helpful measure.

8. *Fluid vibrations—longitudinal or organ pipe*

With the pulsations due to combustion and the number of sharp edges past which the gas flow of the engine passes one would expect longitudinal gas vibrations to occur readily. For reasons not clear to the author these are, in fact, rare; although it is possible that the near sonic velocity and tortuous engine path helps to suppress them.

Plate 9 shows the effect of a longitudinal vibration of the air in a pipe. A characteristic of gas vibrations is that where there is a pressure node there is also a volume antinode, i.e. in a closed vessel a displacement antinode at the exit orifice. This follows from Boyle's law if the pulsations are rapid enough for the compressions and rarefactions to be nearly adiabatic. Hence the lowest natural mode of a pipe closed at one end has a pressure node at the open end with the corresponding antinode at the closed end; while the volume changes are reversed and the displacement is zero at the closed end (which is not really surprising!) and maximum at the open end. The flame of a cigarette lighter, or fishtail burner, makes a very good detector for such air movements as the flame is almost weightless though clearly visible. The curious might note, when performing the experiment with, say, a domestic loudspeaker that the combustion becomes more efficient the greater the movement of the air, i.e. the flame goes bluer.

One does find fluctuations in the gas pressures along the engine in near unstable conditions but these are generally at a frequency fixed by the characteristics of the control system and are often in phase along the whole length of the engine and hence are rather a forced vibration of the gas.

With the pipes carrying engine fuel, and subject to excitation from gear or swash pumps, such longitudinal vibrations do cause resonance effects which are sometimes sufficient to upset the operation of pressure sensing devices. Thus warning lights fed from pressure switches may

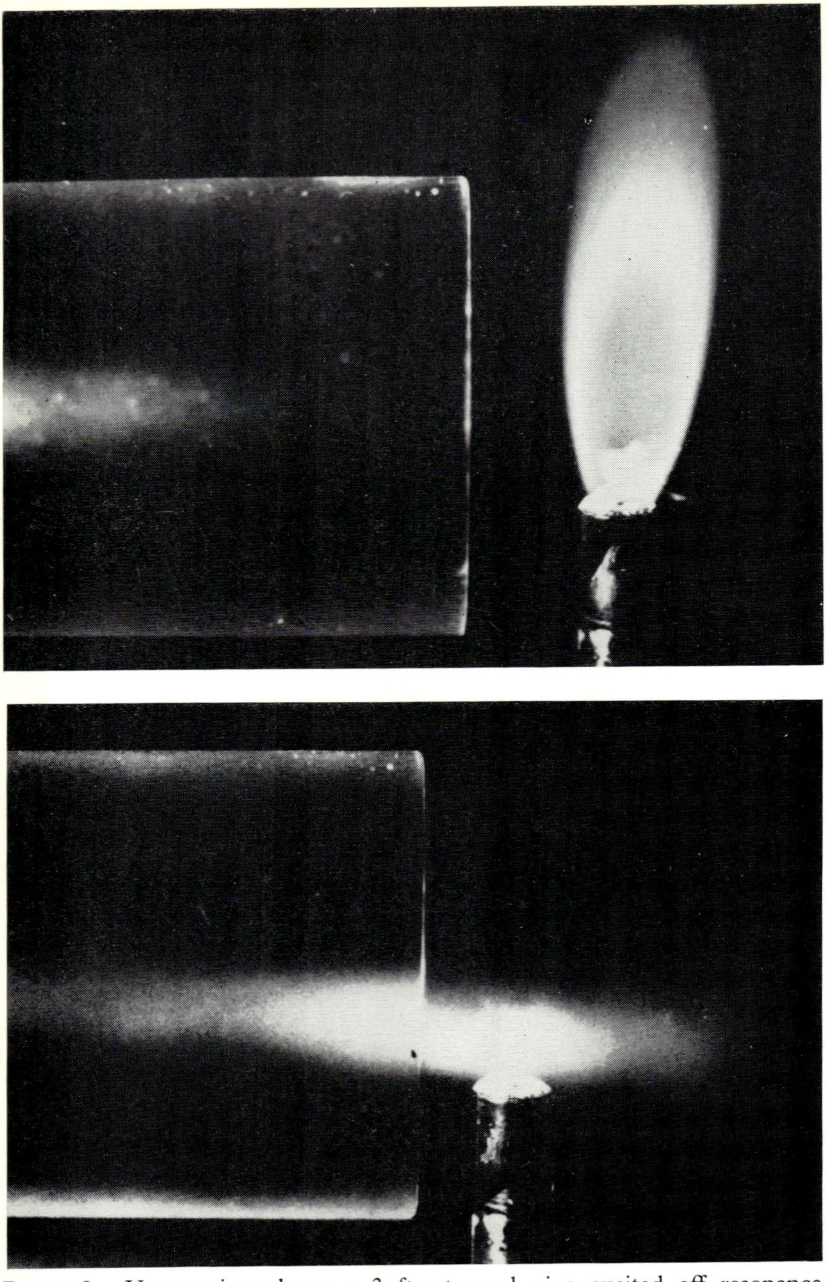

PLATE 9. Upper view shows a 3 ft. stopped pipe excited off resonance. Lower view the same pipe excited at resonance.

flicker (much alarming the pilot) or diaphragm valves operate abnormally. In most of these cases the system is stable to start with due to entrained air surpressing the resonance. When the engine has been operated, and in particular rapidly handled, the air may be eliminated and noticeable resonance occur. In some cases the inclusion of a permanent air bubble in the form of a hollow oil resisting rubber ball of quite small volume will entirely suppress the trouble.

An example of the effect occurred, during the writer's experience, in the low pressure fuel system warning light switch. This effect caused the pilot's warning lamp for low fuel pressure to flicker. The effect was finally determined as a pressure fluctuation, at the frequency of the swash plate pump rotor (26 Hz) which was amplified by resonance with the aircraft pipework and recuperator (2 gal capacity).

The resonance disappeared when the aircraft pipework was removed or if some 20 ml of air was trapped in the system. Without the resonance the pressure fluctuations were of the order of 5 lb per sq. in. (peak to peak) with the resonance they were 50 lb per sq. in.! The air which damped the system was only effective when it was located near the cause, i.e. near the fuel pump inlet.

9. *Fluid vibrations—radial and slosh modes*

The frequencies of resonances in these modes, for typical sizes of pipe, filled with fuel or oil are far too high to be excited by any likely cause. The same is not true for gases and in an early development stage such vibrations can cause much trouble. Such gas vibrations are only lightly damped in a simple tubular construction. However, deliberate breakup of the inside surface with the provision of different path lengths through perforated hollow axial convolutions, is often effective.

CHAPTER 4

forced vibrations

Having dealt in the previous chapters with resonant vibrations of components of engines we now turn to forced vibrations, both of the whole engine and of accessories.

There are three major effects of vibrations, general to the whole engine, which affect respectively the integrity and life of the engine itself; the performance and life of the auxiliaries and accessories mounted on the engine; and the effect of engine vibration on the aircraft it is mounted in.

There are, of course, vibrations passed back to the engine from the airframe but, in general, these are small by comparison with those internally generated by the engine in respect of effect on the latter; since while displacement may be large (several inches) the frequency of airframe vibration is generally below 10 Hz and the accelerations correspondingly very low—except in a crash landing. We now proceed to consider the above three effects of general engine carcass vibration below:—

1. *Vibration of engine carcass itself*

The main sources of this type of vibration in the frequency band below 2000 Hz are unbalance of major rotating assemblies, particularly the compressor–turbine rotor assemblies. Because the responses of the engine mountings and casings are basically nonlinear with respect to amplitude, higher orders of engine unbalance are generally evident; and often significant at second and third engine order. A two or three-shaft engine will have a corresponding number of sets of engine orders, because of the differing speeds of the shaft assemblies.

Unlike the predominantly resonant vibrations of individual parts carcass vibrations are felt all over—though not equally—and are part random and part discrete frequencies of various pieces, often mixed with and modulated by each other. With a shaft output engine there are also large components at reduction gear tooth frequencies. These latter frequencies are often above 2 kHz and, while small in amplitude have very high accelerations.

Inside the engine these vibrations result in increasing the loading on bearings and the fretting of mating parts, sometimes with adverse effects on such components.

If a specific bit of the engine is cracking or fretting excessively it would be dealt with as a case in Chapter 3. If the whole engine vibration is being examined this would normally be done by fitting vibration pickups to suitable places on the carcass—preferably communicating mechanically with the exciting cause.

Thus if engine unbalance is being measured, one would try to site seismic or accelerometer pickups at areas which communicate with the main bearings, while gearbox induced vibration would be sensed by accelerometers as near the gears as possible. See fig. 9 which shows a typical spectrum of engine carcass vibration from 630 Hz upwards, as

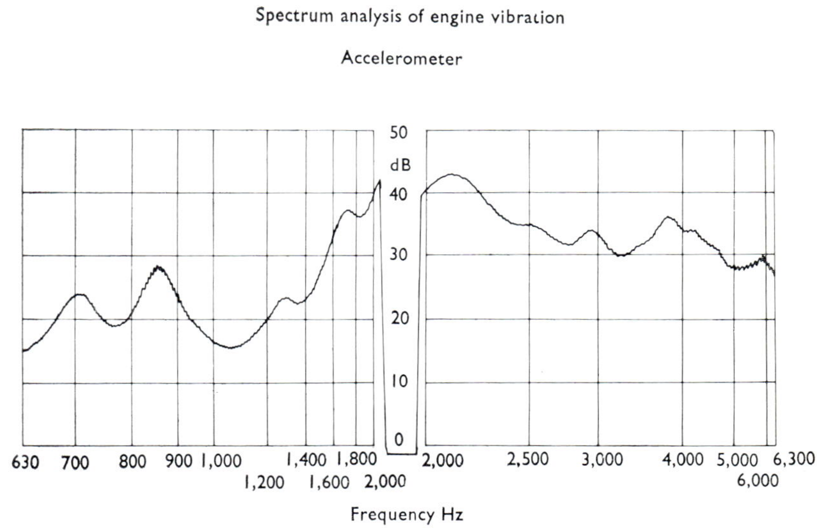

Fig. 9. A typical spectrum of engine carcass vibration from 630 Hz upwards, as measured by an accelerometer fixed to the casing. Note the quite high values of acceleration even above 4000 Hz. The corresponding displacements are of course quite small.

measured by an accelerometer fixed to the engine casing. Note the quite high accelerations even above 4000 Hz. The corresponding displacements are, of course, very small at high frequencies.

2. *Carcass vibration related to accessories*

Where the effects of the vibrations of the engine on accessories are being examined it is general to fit accelerometers at the junction points between engine and accessory, preferably reading in three orthogonal planes. In addition if a piece of the accessory is failing or malfunctioning additional pickups would be mounted at these places; or they would be straingauged.

While the force transmitted between engine and accessory may well be important this can only be found in one of two ways. The direct way would be to interpose force cells at the mounting points, but this is generally impracticable because it would alter the conditions, and the indirect method of straingauging a suitable part of the accessory mounting lug or bracket is generally used.

Where accessory malfunction is being effected by engine vibration additional measurements would be made; such as with displacement or proximity probes, or pressure pickups sensing the affected parameter.

3. *Vibration of airframe by engine*

This is the most contentious problem of all carcass vibration measurements. It is possible to place suitable three-axis seismic or piezoelectric pickups at the engine mounting points, and work on test bed or hangar; but this will inevitably give different results from readings in the aircraft. This arises because the mounting stiffnesses of test bed and aircraft are different. Nevertheless the cost and difficulty of aircraft testing demand that all possible work be done on the bed.

Because the aircraft structure is such that low frequencies under, say, 10 Hz are readily transmitted, the low frequency cut-off used on the test bed in assessing the effect of an engine's vibration on itself (generally just below first engine order idling) is inadmissible and it will be vital to use pickups and read-out equipment going down to zero frequency, or near. An exception to the above is when it is evident that the complaint from the airline is of 'high-frequency vibration'. Experience suggests that this phrase used by a pilot means anything from 30 Hz up to several kHz. At frequencies above about 2 kHz the attenuation in the airframe structure is usually severe and any complaint would probably refer to airborne sound.

Because of local panel and other structural resonances in the airframe, often quite normal amounts of engine vibration lead to complaints which are difficult to verify since they may well depend on the mating of a particular engine and airframe and sometimes on pilot too! An engine which is perfectly normal on the test bed, or even in altitude simulated test tunnel may give trouble when in the aircraft and the aircraft vibration undeniably ceases when the engine stops, so *ipso facto* the engine is the cause.

All measurements of carcass vibration yield results containing a large number of frequencies, some related integrally or otherwise to engine speed and some at fixed frequencies and the analysis of the results is tedious and complex. This will be dealt with more fully in the Measurement Section of the book, but will almost always imply frequency spectrum analysis.

CHAPTER 5

coupled vibrations

Coupled vibrations are those which occur when two separate parts of a system, each capable of resonant vibration, interact with each other. While all vibrations forced or resonant of one part are more or less coupled to similar vibrations in interconnected parts; in this sense the vibrations of various blades in the same stage of a compressor are so coupled. However, the term is reserved in the present context to coupling between two systems each complete in itself.

Such systems in aircraft and engines are where an airscrew is coupled to an engine, an engine to a wing, or an external gearbox to an engine. In everyday life the vocal chords are coupled to the cavities of the mouth, throat and nose which can be separately modified by the position of tongue and palate. In music the lips and vocal organs of the brass wind instrument players are closely coupled to the air column of the instrument and the same is true to a reduced extent with the woodwind instruments.

The broad effects of such coupling vary with the degree of mutual interaction, from the case where the coupling is very small (and both systems continue to display their own separate resonance effects) with little modification of each by the other; to cases where the coupling is very tight and the separate behaviour of the two systems virtually disappears in favour of a joint behaviour, and a quite different set of resonances.

The coupling between an aircraft engine and its wing or fuselage is generally a flexible design and there is little interaction, especially as the resonances of the two parts of the system are widely separated in frequency. Coupling between a propeller and its engine is strong and the separate resonances are close, thus enhancing any such effects. Coupling between an engine and an external gearbox may be weak or strong according to whether resilient couplings are provided or the drive is relatively rigid.

Another manifestation of coupling occurs when two vibration generators, e.g. the two spools of a jet engine, or two engines coupled together at the casings, are at nearly the same speed. In this case ' beats ' occur, which result in the progressive waxing and waning of the vibrations as the two generators cancel and reinforce one another. At the maximum condition the total vibration may be doubled and, if this happens to cause resonance of the structure at a low frequency, multiplied many more times.

1. *Engine and airscrew*

Generally, when a turbo propeller engine is being developed, in association with a specific aircraft, extended tests are made on a test bed with the propeller to be used; and indications of trouble due to coupling between the two would be evident at an early stage in the work. Now while it is possible for the engine to excite the propeller, cases of this are rare, but the reverse is not necessarily the case!

Vibration of the propeller will take two forms; firstly propeller unbalance will force the engine at a frequency which is quite low relative to engine susceptibility—though accessories may be affected, and this belongs to the last chapter; secondly there are marked resonances of the propeller blades which form vibrating systems in themselves. Like other blades they are capable of flap, torsional and edgewise modes easily excited by aerodynamic disturbances.

In the design of the propeller the most harmful of these resonances will be eliminated because of possible failure of the propeller itself. Since the blades are capable of being self-excited by lift changes, vortex shedding and similar aerodynamic forces, these possibilities should have been designed below a dangerous fatigue level as far as the propeller itself goes; but this would not necessarily ensure that they might not be magnified by interaction with the engine so as to harm either the propeller or the engine.

Because it is usual to make tests on the two parts of the system together when an engine is being designed with a specific aircraft in mind, trouble after the engine/propeller are in service is rare.

There is one source of propeller excitation, however, which is difficult to check in a test bed and this is the effect of resonant excitation of the blades when they pass in close proximity to the aircraft structure in cross winds on the ground. There is also the possibility that similar effects occur in flight at unusual aircraft attitudes. It is notoriously difficult to simulate these effects on the test bed, because this cannot be swivelled round in relation to prevailing winds, nor can the latter be controlled at will. The problem is usually dealt with by calculation, as far as the propeller is concerned. When an established engine is designed into a new aircraft, and/or with a new propeller, there will not generally be as much money to spare for extensive engine testing with the propeller—after all the engine is perhaps well proved in a dozen different aircraft and propeller effects are, typically, at frequencies below engine orders and with low accelerations and should not cause trouble.

There was a case, with which the writer was associated, where the above conditions happened to result in a marked effect on the engine which depended on a long chain of coincidences. It forms a useful example of what may happen when everything goes against you, though to check out all such possibilities beforehand may be quite impracticable.

The trouble arose when an established turboprop engine was fitted to a new aircraft designed by a company overseas. Because of the design conditions (1) a larger propeller was required than the engine had carried before, (2) for the same reason the blades of the propeller passed nearer than usual to the fuselage. Because the airline operating the first aircraft was not of long standing the ground personnel were unaware that (3) ground running in a cross wind gives strong excitation to the propeller as each blade passes the quieter air in the shade of the fuselage, with possibility of resonance of the blades themselves.

Because the engine had been in use for a long time in many different types of aircraft, no one thought to check (4) whether such an excitation of the blades would harm the engine, and had they so thought would, no doubt, have decided that if the propeller blades would stand it so would the engine. In any case they also would have expected the ground crews to avoid running in cross winds.

In the event it proved that the third flap of the propeller blades was excited by passing the fuselage wake at a speed within the range usually used for setting controls (and also for taxying). The next unfortunate coincidence (5) was that the engine propeller system had a transverse free-free mode which resonated with the propeller blade excitation. Even these coincidences would not have mattered had not (6) the deflection shape of this mode resulted in a node at the bearing against the turbine disc and an antinode at the bearing at the other end. The effect of the forced oscillation of the free end of the turbine shaft while the turbine discs attempted to remain in one plane, due to gyroscopic effects, resulted in fatigue cracking of the turbine shaft.

This example shows how complicated vibration investigation can become, it required six different effects to combine to produce the defect in the engine! Fortunately such a combination of adverse circumstances is very rare. Naturally the investigation was a difficult one and required the close collaboration of the several firms concerned and academic assistance for its solution.

It is possible to predict some of the effects and, at least, the likely behaviour by measuring or calculating the relations of dynamic stiffness or impedance of engine and propeller over a range of frequencies and then obtain the combined response by superimposition of reversed behaviour curves. The calculations, even by computer, are complicated and the measurement involves sophisticated and expensive equipment so that generally such exercises are not attempted, unless a very long lead time is expected between design and realisation.

The above example was the reason for the exercise illustrated in plate 10 and fig. 10. Albeit the problem had been solved before this experiment was done; the test was made to determine whether such a test would have predicted what actually happened. The results, when the engine tests had been combined with computer calculations for the propeller, was certainly to confirm the coupled resonance which did occur; however, other coupled resonances were also predicted which did not show any adverse effect in practice.

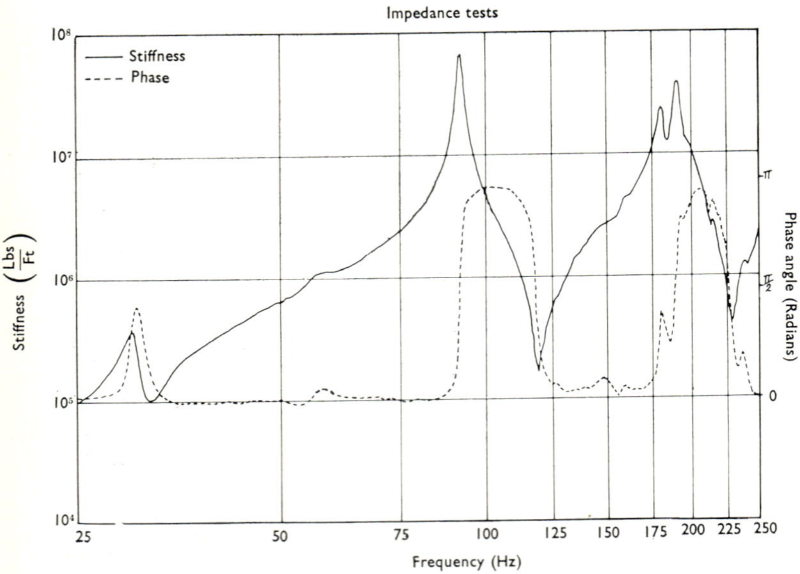

Fig. 10. An actual plot of stiffness against frequency (full line) and of phase angle between force and acceleration against frequency. Note the sharp changes in phase angle at resonance, at 30 Hz and 120 Hz, and also at the anti-resonance at 90 Hz and 190 Hz.

The difficulty then, in this type of test, is that those coupled resonances which may occur can certainly be predicted in advance; but not whether there will be significant excitation to make them harmful. Thus for a new engine-propeller-aircraft combination such methods can only serve to sound a note of caution for the future—not to substitute for tests on the combination which will be flown, when this is available.

2. *Engine and power offtakes*

This section deals with those cases where the engine and additional equipment form a significantly coupled system and not the many other cases where relatively small accessories are forced by the engine to conform to its own pattern of vibration. In respect of the local effects on the engine near the attachment, even the latter may have to be considered as a case of coupled vibration, but generally the engine is so much the major partner as to swamp such interaction. Thus fuel pumps, generators and the like may affect the gear train from which they are driven but do little to alter the main vibration of the carcass.

However, when large amounts of power are taken off; such as to drive an auxilliary gearbox to provide many specialised aircraft services,

PLATE 10. An example of a setup for measuring the mechanical impedance, receptance or stiffness of an engine over a frequency range. The engine is shaken by a vibrator in the plane of the propeller attachment. A force cell in this drive measures the force and an accelerometer at the top records the resulting motion.

or a helicopter rotor; the combination forms a fairly tightly coupled system and neither can safely be considered by itself. Because there will be in almost all cases, flexible couplings to avoid bending moments developing between the two units, the effects are mainly evidenced in torsional coupling. It is clear that in order to transmit significantly large torques the compliance (flexibility) in a torsional mode must be low while such gear boxes often drive units with large polar inertias.

As noted in the preceding section impedance plots can be made of each component and the results married on paper, but this requires great care in adequately simulating the mutual anchorage coupling between engine and gearbox. It is therefore more common to undertake such testing on an aircraft itself since fuselage stiffness is frequently a major factor.

Such tests also require that all the many loads fed from such a gearbox be correctly applied at various engine speeds and powers and the whole exercise may be quite protracted. Such torsional effects have to be assessed in respect of rotating shafts which calls for torquemeter devices on these shafts capable of operating over a range from, say, 10 Hz to 3000 Hz. It may be necessary also to measure the effects on the engine and gearbox frames with suitable vibration pickups.

In this type of investigation a major factor is the likelihood of sudden changes of load from the auxiliaries on the gearbox, since these are frequently responsible for the highest excitations. A sudden offloading of a generator, for example, because a circuit breaker trips, may set up a damped train of torsional vibrations of large value; also the cyclic pattern of loading due to a voltage regulator may well coincide in frequency with a major torsional resonance of the whole system.

One of the main difficulties in such testing is getting adequate information on the various transient changes of loading which might occur in service. The average operator, unless specially primed and very observant, will not think to mention such events and every often one of the investigating team may have to spend some time with the operator noticing all that happens; before the actual tests are programmed.

It may well happen that some of the important effects take place while the engine is being started and the various auxilliaries come on load. Constant speed gearboxes too are a fruitful source of torsional forms of excitation. Because one never knows when something interesting will appear it is usual to tape record (multichannel) all the likely parameters to avoid shooting miles of film or recording paper, on most of which there will be nothing to see!

3. *Engine and airframe*

The complaints by customers which spark off testing in respect of this form of coupled vibration arise in many ways. Pilots may complain of seat vibration at low frequency or of ' pins and needles ' in their hands when on the throttle. They may note vibration of gauge needles, or the glasses of the gauges. Passengers may comment adversely on noise or seat vibration.

All these complaints merit serious consideration in selling an aircraft or maintaining its commercial viability. Sometimes the effects occur on the ground when undertaking maintenance and control checking and sometimes in the air. Now there are many other items in a large aircraft which can cause vibration apart from the engine, e.g. cabin air conditioning and pressurizing, hydraulic generators for undercarriage and other aircraft functions; also aerodynamic roughness around the structure when in flight and so on. Since aircraft are rarely flown without the engines running, naturally the engine first gets the blame!

The attachment of the engine to the aircraft is, necessarily, based first on the need to transfer the thrust from engine to aircraft wing, or fuselage, and secondly to deal with the g loadings (accelerations) between the two under realistic conditions of use. The need to transmit such large forces safely implies strong but flexible coupling from engine to airframe.

Fortunately the natural frequencies of the airframe are generally a decade lower than those of the engine excitation and heavy coupled resonance of large areas of the aircraft structure with the engine are few. However, the out-of-balance forces from the engine can travel along spars, bulkheads, deck and control levers and if they reach a local panel which resonates at the same frequency quite a large amplitude can build up locally. Such a local resonance may be felt if of low frequency, or it may be heard as re-radiated sound, if at engine excitation frequencies.

However harmless, from the point of view of failure of components, the above effects are they militate against the comfort of passengers and the efficiency of the crew and must be regarded seriously. Doubtless the writer will incur some adverse comment from airframe designers, but it is his opinion that the suppression of such effects is primarily in the sphere of the aircraft design and not that of the engine manufacturer. The provision of suitable damping on large flat panels and of vibration isolators in control rods is a much cheaper alternative than applying similar measures to—quite hot—parts of the engine; and as far as the noise re-radiated into the fuselage from its panels, this can be suppressed to any desired degree by the use of suitable damping in upholstery and carpets much more easily than it can be eliminated at the engine.

The engine manufacturer's job, in the writer's view, should be to reduce the level of vibration generated within so that it does not cause failure or malfunction of any engine or associated auxiliary. If this be done effectively then engine carcass vibration should be no more than can be dealt with by standard noise and vibration reduction techniques within the aircraft. The latter would be needed, even if there were no engines, to deal with the effects of conditioning of cabin air and aircraft control systems themselves and aerodynamic sources of vibration not due to the engines.

If it is necessary to carry out investigations, these are normally made with accelerometers, seismic pickups and microphones mounted at the critical points and their use is dealt with in the second half of the book.

4. *Coupled beats*

When two vibrations of nearly the same frequency are fed into the same structure, interaction between them causes a phenomenon generally referred to as ' beats '. This is a periodic waxing and waning of the sum of the two vibrations, which has a characteristic sound and feel. The effect is easy to hear when a multi-engined propeller aircraft flies by, if the engines are slightly off the same speed, when it sounds as a throb, or pulsation, at a frequency of perhaps 2–10 Hz—all too familiar to the older reader who lived through the air raids of the last war!

A musician is very familiar with the effect, and a tuner uses it to set up the equal temperament scale on a keyboard instrument such as piano or organ. Octaves are tuned in unison but other intervals are made to beat, so as to be slightly incorrect in all keys, rather than correct in one and intolerable in others. A ' fifth ', for instance, is tuned to about 4 beats per sec, i.e. a frequency error of 4 Hz from the 3:2 ratio of a perfect fifth.

Engineers frequently notice the same effect in the floor vibrations of machine shops. This behaviour, by itself, is not generally of consequence, unless the beat frequency happens to cause sympathetic resonance of parts of the structure.

The author does recall an instance when the vibrations from two pedestal grinders did make it impossible to use a particular optical comparator in the nearby standards room. The cause took a lot of tracking down, because it was only occasionally evident—always on Friday! The two grinders, on which occasional jobs only were done, were at opposite ends of the toolroom while the standards room was next door. Most of the time there was no interference, but if two machine men happened to walk up to the grinders at the same time, and if the grinding wheels were badly out of balance (they were dressed each week on Monday, so Friday was their worst day) they could be running at slightly different speeds. If this difference in speed coincided with the resonance of the suspension of the optical system of the comparator this was set into such a state of vibration as to be quite unusable. As may be imagined it was not easy to track down this vibration trouble, nor indeed to convince a hardheaded foreman that it did really happen that way!

Some aircraft have their engines fitted with automatic synchronizing control, so that beats are less likely to occur, rather to get rid of the unpleasant effect on the passengers than because of any real harm to the aircraft. Also where, for instance, pairs of engines are grouped close together, and fairly rigidly coupled, as is not uncommon in some industrial installations where jet engines are used as gas producers, the beats between engines when close in speed can trip supervisory vibration control gear—if this is set too close to the bone.

CHAPTER 6

typical measuring devices

Before proceeding to a detailed discussion of the individual applications of measuring devices it is worth while spending some time discussing the general considerations involved.

It is conventional to apply the term ' transducer ' to devices which are used for changing one physical effect into another—as faithfully as possible—either for control of a process or operation, or for measurements on the same. From this it might seem that the same devices would be equally suited in both fields but this is not always the case.

For control purposes it is often preferable to transduce a measurement into fluidic (or hydraulic) terms since these are cheap and easy to apply to equipment being controlled, but for measurement purposes it is almost universal to transfer, or transduce, into electrical voltages or currents. This is because in control applications the information is applied directly to modify the operation but for measurement it is usually necessary to record, in some manner, what is happening.

In addition the amplification, over quite wide ranges of magnitude, is easy electrically and there are no practical limits to the speed of operation or the remoteness of the equipment. There is also the consideration that processing of the signal, e.g. by integration, differentiation or selective filtering—together with the ability to make intermediate recordings on magnetic tape and to present the information in analogue, or digital form—all imply electrical methods.

It is unfortunately true that the measurement of any effect necessitates some major or minor change in the quantity being measured, and it is part of the art of measurement engineering to reduce this to a minimum. It is also the case that almost all practical transducers fail to be completely single-minded, i.e. they respond somewhat to side effects to which, ideally, they should be blind. One of the most common causes of such failings is the effect of temperature changes when transducing other parameters and nearly as inimical is vibration when this is not being intentionally measured.

Logically one would expect that the direct measurement of the required parameter by a transducer would always be the best, but in practice it often happens that one chooses an indirect effect because it is less sensitive to side-effects than a direct one. At other times the inaccessibility of the location to direct observation or the insensitivity of the direct method, dictate the use of an indirect measurement.

For example it is very difficult to measure directly the fluctuating skin stress in a piece of metal of irregular form but the strain associated with such stress can be measured by a resistance straingauge—even if the part be rotating at high speed and at an elevated temperature.

PLATE 11. A part of a recording caravan equipped to deal with multi-channel tape and galvanometer records of straingauges, pickups and other transducers.

Again the temperature of a hot object may be measured directly by an attached thermocouple if the object is fairly massive; but a very thin sheet might have its own temperature modified by heat flow down the thermocouple and an optical method might be more suitable.

When considering indirect transducers it is especially important to select those insensitive to side effects. For instance, to cull an example from the writer's past experience, it was required to measure the dimensional change of a ' Tufnol ' cylinder about $\frac{1}{2}$ in. in diameter when heated, dry or wet, from room temperature to 150°C. Since it was necessary to use an oven it was difficult to measure the diameter by conventional means *in situ,* and suggestions were made that capacity transducers be used. Unfortunately these would have been hopelessly invalidated by the changes in dielectric properties of the material and the ambient air with temperature and moisture content. Eventually a modification of the N.P.L. level comparator was used with the beam from the mirror passing up the chimney of the oven on to a scale above.

Another case where indirect means are attractively deceptive is in using thermocouples to deduce length changes in jet pipes and the like. The catch here is that one would need an enormous number of couples to give a true mean measurement. In such a case it is better to use linear transducers, however difficult it is to keep them cool—or if the changes are large enough a foot rule!

Another general consideration in the use of transducers might usefully be raised and that is the choice, when it exists, between direct amplitude output transducers and frequency modulation ones. With an amplitude sensitive device the greatest care has to be taken to ensure accurate and unvarying amplification. However, a frequency modulation system (f.m. for short) does not suffer so much from this defect and for this reason is generally preferred; while when tape recording is employed it is essential if the datum is to be preserved, together with response below 30 Hz. On the other hand, the carrier frequency used with f.m. equipment puts a top limit on the rate of change which can be measured and for this reason very high frequency effects need direct transduction. These points are elaborated in later chapters.

The effects made use of in changing a mechanical quantity into a corresponding electrical one are many and varied. One of the commonest is the change in resistance when a wire or foil of high resistance material is stretched. The piezoelectric effect, which makes certain natural or artificial crystals produce a voltage between their surfaces when they are deformed, is also widely used. Thermoelectric voltages produced when a junction between dissimilar metals is heated are used for temperature measurements, while the sensitivity to light of other special materials is widely used for speed measurements.

The changes in resistance of metals when they are compressed are used for direct pressure measurements. Changes of inductance or capacitance when coils carrying an alternating current, or plates fed with an alternating or direct voltage, move relative to each other are used for length and pressure measurement. In fact there is hardly an effect known to physics which is not used at some time or other for transduction of mechanical engineering effects into their electrical equivalents (or analogues).

For process control it is becoming more common nowadays to use mechanical analogue to electrical digital transducers. However, for transient measurements of rapidly changing or high frequency effects, the inherent difficulty of securing a sufficiently rapid response militates against digital devices in many cases. After the transducer itself though, once the signals have been electrically processed (filtered, integrated and amplified, etc.) a conversion to digital output may be necessary in order to present the information to a computer for mathematical processing or evaluation.

No apology need be made for the very large section of the book devoted hereafter to specific transducers, their choice and application, since clearly unless the primary transfer from mechanical to electrical effect is correctly made, all else is useless. In fact the transducer is the Achilles heel of the mechanical measurement engineer, who sometimes envies his counterpart in electrical engineering where the quantities to be dealt with are already in the preferred form.

All practical transducers give faithful transduction over a quite limited range of the parameter involved and are affected by environmental conditions to a greater or lesser extent. These factors make a choice of the most suitable transducer onerous, and puts a premium on wide experience in the field for a successful measurement job. Another important consideration is that the modern use of tape recording means that, if necessary, the information from a given test may be successfully replayed into a wide range of equipment for processing the results, but the validity of these depends entirely on the accuracy of the original information given by the transducers.

There are particular considerations affecting the measurement of rapidly changing, non-cyclic or other random effects. For example, if the change of temperature with time is being measured no power on earth can improve the results if the temperature was changing at a rate many times greater than the response time of the thermocouple, used for the measurement, permits it to follow. Similarly if a pressure was read by a transducer fed from, say, a 3 kHz carrier it is no use the engineer interested in the results coming along afterwards and asking if any effects could be found at, perhaps, 2·5 kHz—this information is lost for ever.

Correspondingly, if say rotational speed is being measured while it is changing rapidly; one can measure the speed accurately over an indeterminate time or the speed inaccurately at a precise time. The degree of accuracy which can be obtained depends on the number of ‘ bits ’ of information provided per revolution. Thus if the speed–time curve of an engine start is required accurately one may need a 100-tooth wheel on the shaft in order to record the slower part of the curve. However, if a high speed is later reached the output of this wheel may be too high in frequency for the equipment to follow.

The above considerations lend especial force to the importance of the ‘ customer ’ approaching the measurement engineer explaining the problem he is trying to solve rather than saying—in effect, and sometimes bluntly in these terms—“ Put me some straingauges here and here and some thermocouples there and there.” If he does this he will get the results he deserves—but seldom those he needs. This book will have failed in its purpose if it is taken as a means whereby a design or project engineer may feel that any Tom, Dick or Harry can undertake satisfactorily an important instrumented test involving many parameters, and perhaps impossible to repeat if the results are unsatisfactory.

The writer would also point out that whereas often the ' customer ' is concerned, and rightly so, when results do not fit a particular theory he should logically be equally worried when they do fit! A properly staffed and equipped Measurement Facility is worth its weight in gold and should be treated accordingly. If results are unpopular it is just as important to review the conditions of build and test of the mechanism as the integrity of the instrumentation and processing.

It should not be necessary to observe that no more than one variable should be changed at a time, unfortunately the writer's experience has often shown a complete disregard of this fundamental step in scientific experiment. An example may make things clearer; if it is desired to measure the changes in temperature of the cooling air to a bearing with changes in flow, it should be clear that special steps will have to be taken to ensure that the thermocouple reading of the gas temperature must remain independent of gas velocity. Unless the measurement engineer is fully informed, it is quite easy to demonstrate that the temperature does not vary at all—or even rises with increasing flow! All that is needed is a badly sited and/or heavy couple with a good soak-back path to the casing, which will recover a greater proportion of the (lower) temperature as the velocity of gas past it rises. Neglect to eliminate radiation from nearby metal can produce even more curious effects. In this case a sheathed couple with suction may well be necessary.

In the following pages the writer has attempted to discuss most of the transducers and recording equipment in general use. Inevitably many of the less usual methods have been omitted from lack of space. The references above to transient effects are made because it often happens that the vibrations being investigated are dependent on such changes in the engine. While not strictly measurements of vibration these parameters have to be recorded in order to understand the causes of the vibration being investigated. The investigation of transients in the control systems of aero engines, and other machinery, is a field in itself and outside the scope of the present book.

It is worth noting that more often than not the Instrumentation Engineer engaged in vibration measurement, was not trained as such; the breed is still somewhat a *rara avis*. This particular bird is a hybrid of varied parentage! He may be the offspring of a physicist, electronics engineer or electrical engineer, married to a mechanical environment, vice versa the product of a mechanical engineer and an interest in music, acoustics and electronics.

The present writer falls into the latter category, having been a mechanical engineer with musical, acoustic and electronic interests for 20 years before he did his first vibration investigations. Whatever his background, the successful vibration engineer will have to become familiar with a wide variety of disciplines; operating equally happily in the fields of electronics, acoustics, optics and most other areas of physics. Yet, at the same time, he must never lose sight of the engineering purpose behind the investigation.

There is always the temptation to be charmed by a novel or spectacular method rather than the humdrum one, but careful consideration should always be given to possible side effects, or loss of accuracy, inherent in the new method. At the same time there is nothing sacred in old methods, and instrumentation engineers must be up to date in all the new techniques, and as ready to use radio-active components with ionization chambers, or Doppler radio equipment, as they are to use their eyes and ears.

CHAPTER 7

specific transducers and their applications

THIS chapter deals, in moderate depth, with the properties and limitations of various of the more common transducers used in engineering measurements. The list is not complete by any means, but those dealt with will cover the most usual cases met with in engine and rig testing. The writer could not devise any logical order in which to select particular types, since the basic physical effects used are not themselves necessarily related.

In addition to the above dilemma, the consideration arises that each given type of transducer can be used for indirect measurements in other fields; sometimes because they are smaller than an otherwise more appropriate type, or because all the channels suitable for one type are already in use, or for other reasons, e.g. because a type may be suitable for telemetry. Such considerations will be explained during Chapters 8 and 9.

The above remarks reinforce the purpose of such a book as this, directed to recently qualified engineers in the art, or those older ones experienced in other fields but not specially expert in measurement of cyclic effects. The most obvious transducer is not always the most suitable.

1.1. *Strainguage characteristics*

As indicated in the previous chapter, straingauges are more often used in order to infer stress than they are, in their own right, as indicators of strain. While they may appear, to the outsider, to be messy to stick on and hence not a very sound engineering method, besides being very vulnerable to damage; their virtues are very real. In fact, their disadvantages are strictly related to their merits.

A straingauge of the resistance type consists of a zigzag of fine wire, or a grid of suitably etched foil, or a filament of semiconducting material, which is cemented to—but insulated from—the part to be measured.

The effect of any strain, i.e. extension or stretch of the surface, is to change the resistance of the gauge; this will change the current if the voltage is fixed, or the potential if the current is constant, or the degree of unbalance of a Wheatstone bridge of which the gauge is an arm. Since practical gauges are thin and weak their stiffening effect on metal components is very low and generally negligible. See fig. 11 for the basic circuits involved.

Since they are normally manufactured to be predominantly sensitive to one axis their use can enable the stress to be calculated quite simply (when the modulus of the material is known) within the limit of proportionality. If strain beyond the limit of proportionality occurs, it is still possible to obtain true stress if the complete stress–strain curve is known—together with the stress history to the time of test.

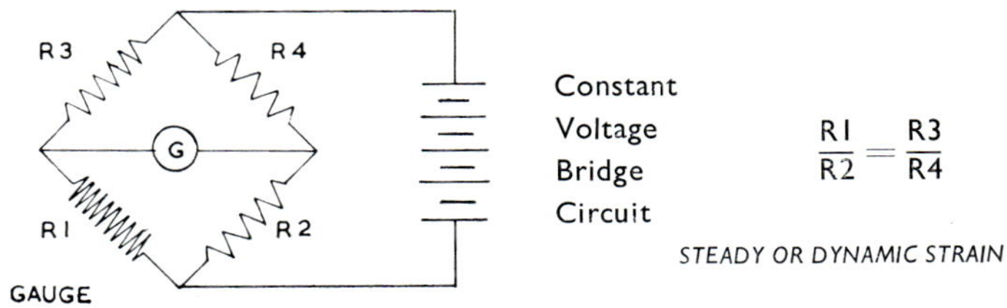

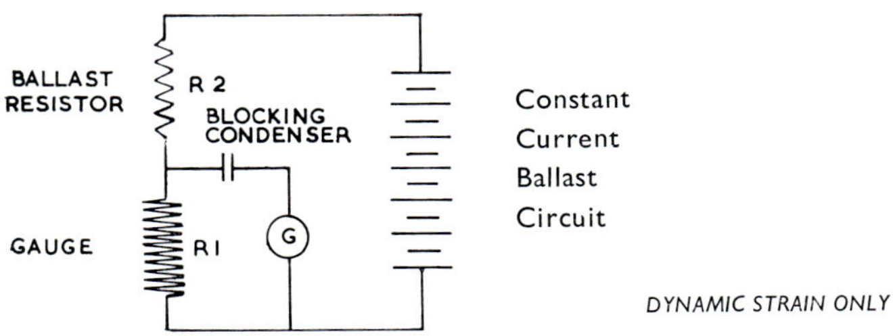

Fig. 11. Circuit 1 requires either power supplies or galvanometers to be isolated from earth and from other gauges. Circuit 2 allows one side of power supplies and one side of recorder circuits of several gauges to be common and earthed.

An enlarged photograph is shown on plate 12 illustrating a selection of common types of gauges. In addition to the normal single-axis gauge many regularly required types of combination gauges are available, such as rosettes, circular wound gauges etc.

Where measurements are required on small local areas, at low or high temperatures and in adverse space or ambient conditions, e.g. on the aerofoils of rotating compressor or turbine blades, there is little present alternative. It is true that tip amplitude and frequency may be measured by stationary probes—either inductive or optical, or even by radio Doppler methods—but these require more complicated equipment and suffer from many side effects, so are only used in special cases.

The strain gauge, in place with its insulation, may be no more than 0·010 in. thick and can be so firmly cemented that it will withstand centrifugal fields of 20,000 *g* with near supersonic air velocity over it. It is a low impedance device and gives no output—from strain—without energization, which means it can be switched on and off at will. This

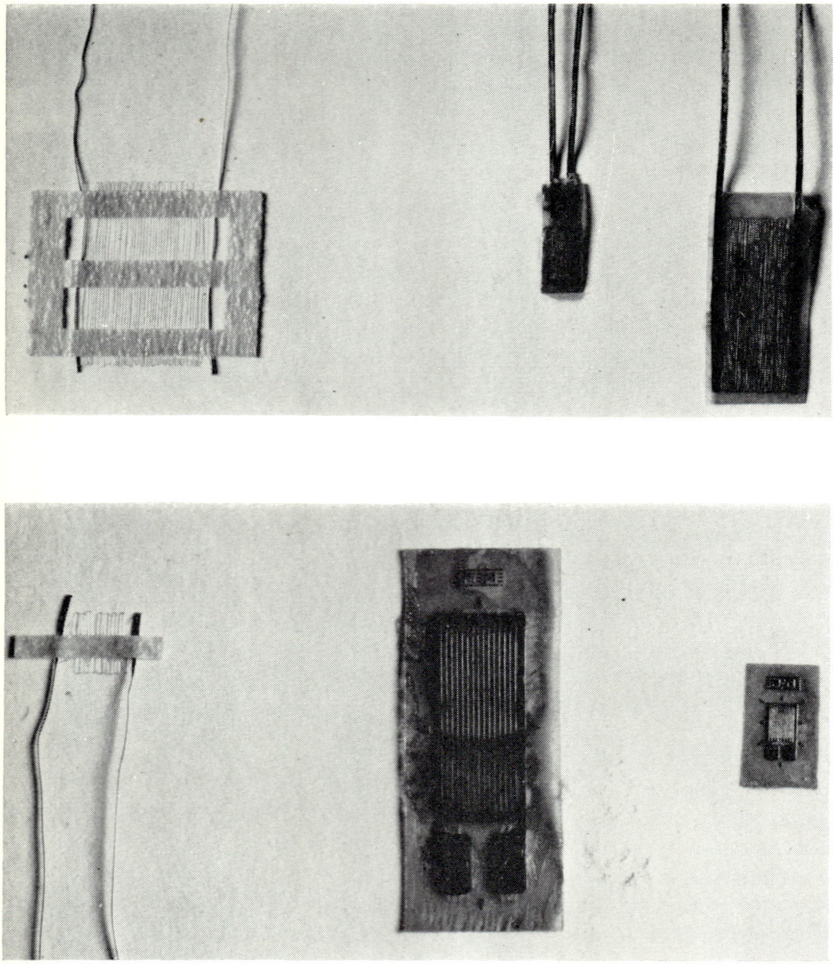

PLATE 12. A selection of straingauges, those on the left are for embedding in Brimor cement, those on bottom right are foil gauges.

characteristic is invaluable when electrical interference is present; since without polarizing no strain readout can occur and any effect that appears must be due to interference, which greatly assists diagnosis when in trouble with a test.

The low impedance, characteristically from 100 Ω to 10 kΩ, makes subsequent amplification easier; without introducing capacity feed-back from such interfering sources as often occur in testing due to the electrical services normal to engine control. Also the gauges can be accurately calibrated statically before and after a test.

As far as size goes the straingauge is easily the smallest of transducers which can generally be applied to actual engine components. Gauges of active length no more than 0·025 in. are readily available; suitable configurations of gauges can be chosen which enable actual components to become transducers of displacement, force and pressure. See plate 13 for a selection of straingauge parts incorporating gauges.

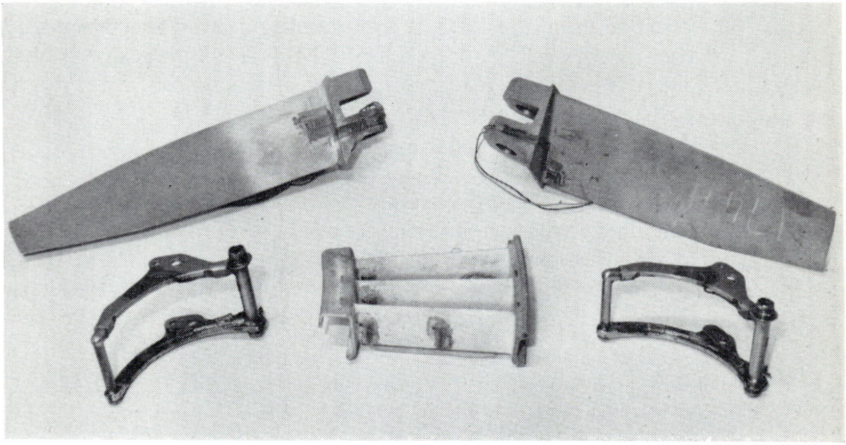

PLATE 13. A selection of straingauges on engine components. Two compressor blades, one with a gauge on the root of the aerofoil and the other with a gauge in a different chordal position together with some stator blades fitted with rather small gauges are shown. Two oil lubrication pipe assemblies are also illustrated, both fitted with two gauges.

When very low strains have to be measured, or when competing sources of electrical noise are high, much greater sensitivity to strain (about 100 times better) can be obtained from semi-conductor gauges; though these are more difficult to handle, and more variable in their calibration, than wire or foil gauges and are only used when necessity commands.

Books on straingauging give full information on the associated problems, and on the many types of gauge available, and the scope of this book is restricted to indicating their use and application in terms of the job they do in retrieving information.

One of the great advantages of resistance straingauges is the very large range of strain over which they give a linear output. This range can easily extend over a ratio of 100 : 1 from top and bottom, and with special care even by another factor of 10. Capacity and inductive methods, on the other hand, are essentially non-linear and to obtain a ratio, between minimum and maximum, of over 10 requires considerable care.

However, as indicated later, when pressure and other transducers are used, where straingauges are the sensing element, the effective range may be limited by hysteresis or lack of proportionality in the mechanical elements of the transducer itself—not the gauge.

In favourable cases when optimum materials are used in load cells, proving rings and pressure transducers, with special attention to the metallic elements, the range may be substantially linear over a range of 100 times.

Since this book is of general scope, and does not attempt to be a primer on straingauging, no details are given relating to cementing gauges, or the associated leads and/or sliprings. There are many dodges used in manipulating and arranging the gauges on actual components, but these are the province of the instrumentation specialist.

1.2. *Straingauge arrangements*

In the academic field and in some rig testing it is generally possible to use a sufficient number of gauges as to fully delineate the stress pattern. In the case of other rig tests, and generally in engine testing, factors such as the geometry of the component, or the number of leads possible, may make it impracticable to use more than one gauge per position.

In such occasions as occur following failure of a component by cracking, the direction of the principal stress is known. In these cases there is no doubt as to the correct disposition of the axis of the straingauge—it is at right angles to the crack. It is also certain that where cracking occurs at an edge there can be no surface stress at the edge other than along it, a point sometimes forgotten. Hence in these circumstances there is no doubt as to the right direction for the gauge.

The question of the ideal length of gauge is often not simple, since it depends on the stress gradient—which may be unknown. Where a crack occurs in part of a component where the stress changes rapidly the best approximation to the maximum skin stress is given by the shortest gauge *which spans all the crack positions*. The reason for the italics is that the writer has experienced cases where validity of the results has been lessened by the use of too small a gauge. The reason being that individual components will vary slightly in the critical area and move the point of maximum stress significantly in relation to an unduly small gauge.

For example, if the radius at an encastered end of a compressor blade is, say, 0·125 in. and cracks have been found at all points of the radius, the use of a very short gauge may be unwise. It may well be better to read a somewhat smaller proportion of the actual maximum stress, but be sure that this will occur under the gauge, than run the risk that the gauge may not cover the maximum. These considerations are particularly important where large clearance pin joints are used, e.g. in compressor blades and wheels. A diagram illustrating the mechanism of the effect is given in fig. 12.

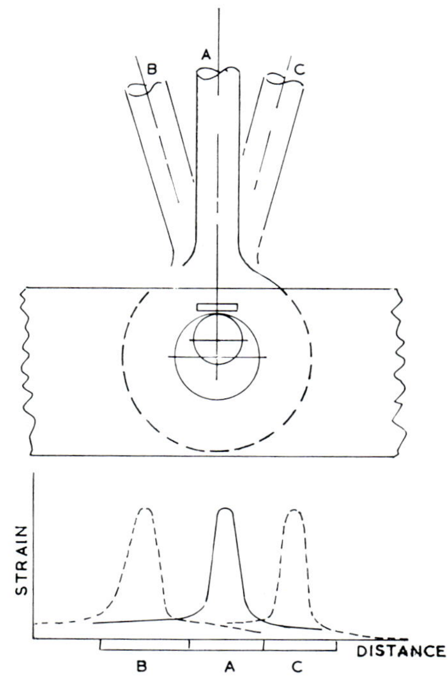

Fig. 12. Short gauge A reads 80% of peak strain A but only 40% of peak strains B and C. Long gauge BAC reads 27% of peak strain whether at A, B or C.

Where it is important to find the magnitude accurately, and not just the engine condition at which resonance occurs, then photoelastic tests may be needed to ' factor ' the gauge reading. In most of these cases it is better to obtain effective co-operation between the stress engineer and the instrument engineer.

It is possible in other cases to arrange several straingauges so that a particular type of loading, e.g. torque of a shaft, will be read while bending and tension—compression are cancelled out; correspondingly either bending or tension—compression can be selected (with different wiring of the same gauges) and the other types of loading neglected.

Such arrangements are illustrated in gauge makers' catalogues and in specialized books on straingauging.

2. *Proximity probes*

These devices generally depend on change of capacity or inductance with relative movement of two parts of the mechanism. As with straingauges, such devices are also often the sensing element of other transducers. While proximity probes are really only a particular case of linear transducers dealt with in the next section, they are so frequently thought of under this name as to merit a separate section. Also, generally they are used by themselves, over very short distances to deduce very small dynamic movements.

2.1. *Capacitative*

Capacitative devices generally depend on making the electrical condenser, formed between a fixed and a moving part, change the frequency of a tuned oscillating circuit, less usually they form part of a capacity bridge.

They are at their most effective over very small movements of, say, up to 0·1 in. and with ease can go down to 0·0001 in. changes or even a decade lower. As flat plates they obey a square law and hence need setting and calibrating near their operating distance. They can be made of greater range, and substantially linear, by using a pair of coaxial sleeves but are then rather large and sensitive to transverse movement, i.e. their cross sensitivity is high. Indeed any form of capacity transducer with only two electrodes is relatively high in cross sensitivity and its successful application is mainly to measuring movement when this is along one axis only.

On the other hand, because of the lack of any magnetic parts and their low inductance, they can be used with very high centre frequencies of the oscillators and therefore can give response to frequencies up to 20 kHz or even higher. Also because they work best in a frequency modulation system they lend themselves more readily to rotating electronic telemetry systems of various types. As mentioned in the first half of the book, the highest signal frequency which can normally be used will be about one-fifth of the centre frequency of the oscillator.

With some f.m. systems the characteristic of the demodulation is an S curve and incorrect adjustment of the capacities of the gauge circuit and/or its length of leads may leave the system operating on the low slope side arms instead of the high slope (high sensitivity) centre section. Even worse results accrue if the system should be set on either nose of the curve when ambiguity occurs giving identical readings for two different separation distances.

This is shown on fig. 13 where the desired operating area is from B to C (actually, say three-quarters of this is used) while a range either side of B or C gives ambiguity and operating A to B or C to D gives low sensitivity. It must also be remembered that, if the ambient temperature or dielectric properties of the fluid or gas between the active electrodes change, the room temperature calibration may shift so much as to result in operation on an undesired part of the curve.

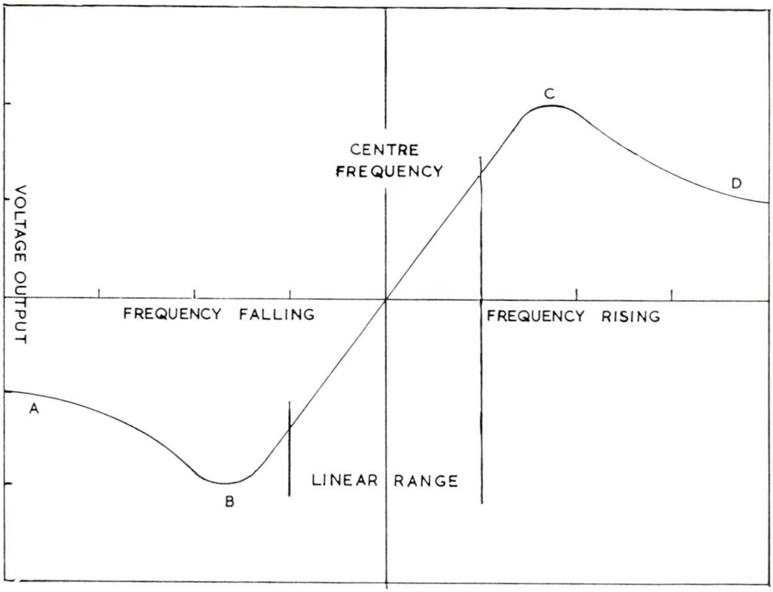

Fig. 13. Behaviour of a discrimator or demodulator in an f.m. system.

Change in capacity of the leads with temperature may result in similar effects. While changes in padding capacity may effect a cure it is vital to ensure that calibration checks cover all operating conditions before recording an engine test.

One of the types of measurement where capacity probes are more suitable than most others is at high temperatures—800°, 900° or even 1100°C are quite practicable, using suitable metals for the electrodes and sintered oxide for insulators. Generally also the engine part can be one of the electrodes, and since it is desirable that one electrode should be several times the area of the other (to reduce cross sensitivity) this makes the method ideal for measuring of movement of sheet metal parts at the hot end of an engine. However, it is important to realize that relative movement is what is sensed and therefore any undesired movement of the (nominally) fixed electrode will be interpreted as movement of the part. Sometimes dynamic resonances of engine parts have turned out to be resonance of the bracket supporting the 'fixed' electrode so that this possibility must be checked by vibration tests before use. Since the system preserves the true datum it gives steady state results as well as dynamic, but drift can be caused, again, if warping of the support of the ' fixed ' electrode occurs. Nevertheless, for awkward places and conditions this type of measurement may be a very convenient method, if not the only one possible. Where the distance between a rotating and fixed part is being measured, the calibration under operating conditions may be displayed by making a ' step ' or ' steps ' on the rotating part; which will scan past the fixed electrode and give a step change in the output of known amount. This also, incidentally, provides a shaft order impulse which may be needed for other purposes.

Standard systems of oscillator and discriminator (or demodulator) are available with a wide range of centre frequencies.

2.2. *Inductive*

Inductive proximity probes can be made to have a long or short usable range, and will generally behave rather more linearily than capacity devices but, because they involve wound coils, are more limited in temperature. As a stock item 100°C is about the limit, though they can be made with ceramic formers and glass insulated nickel-plated copper wire to work as high as 600°C. In the lower temperature range versions they are often used with an oscillator frequency under 10 kHz while at the higher temperatures—where ferrite cores are useless—1 MHz or higher.

Like capacity probes, unless used with guard shields, they are affected by movement of metal parts around them—they do not function well buried in metal, particularly low resistance metals, where the flux is robbed from the active gap by undesired eddycurrents. Cross-sensitivity may be made lower than with capacitance types, since the field from the probe coil can be markedly directional. The same remarks relate to working on the correct part of the response curve as were made in the previous section. Lead wire capacitance is less bothersome with inductive types though it cannot be neglected. Plate 14 shows an example of an inductive probe.

2.3. *Other types*

Coming into use, but not yet in quantity, are other types of probes. For instance a Gunn diode, used in a suitable wave guide cavity and operating in the region of 13 GHz can be made to work as a sensitive indicator of dynamic movement. Originally developed as a velocity measuring device depending on the Doppler effect, they can detect oscillating movement—by reason of their sensitivity to the corresponding velocity changes—well into the MHz range—and can ' look ' through a nonmetallic window ; so that movements of, say, 0·010 in. can be read at distances of several feet. One must remember that movement of the device itself comes into the measurement so that brackets must be well designed and quite free of resonances in the sensing range. The whole device can, if necessary, be mounted in an air cooled box and ' look out ' through the exit air quite unaffected thereby.

Datum changes are not recorded at all. Photocells and photodiodes, particularly the photo-voltaic silicon type, as used in solar cells, can be

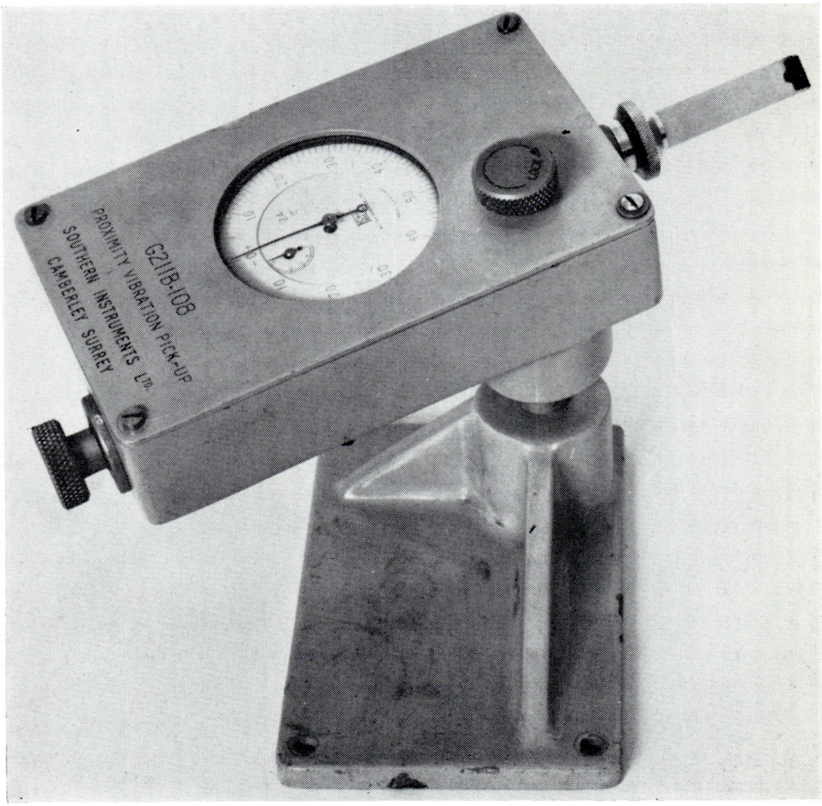

PLATE 14. An inductive proximity probe, the active part being the small coil embedded in the black moulding at the end of the blade. The central carrier frequency for which this is wound is 2 MHz.

used by arranging that movement of the part being measured alters the amount of light from a source, but are not often a satisfactory method except in conjunction with a fibreoptic light guide; when they can reach otherwise inaccessible places. Plate 15 shows a commercial speed radar which will also measure dynamic displacements.

3. *Displacement transducers*

Displacement transducers, linear and angular, are many and varied, both in the manner of their application (dealt with in Chapter 8) and in principle of operation. Almost any physical effect can be brought into play to transduce a steady-state movement but the field is rather more restricted when it is necessary to determine dynamic movement. Since this book is dealing with the latter, the information given neglects such methods as are inapplicable for cyclicly varying measurements.

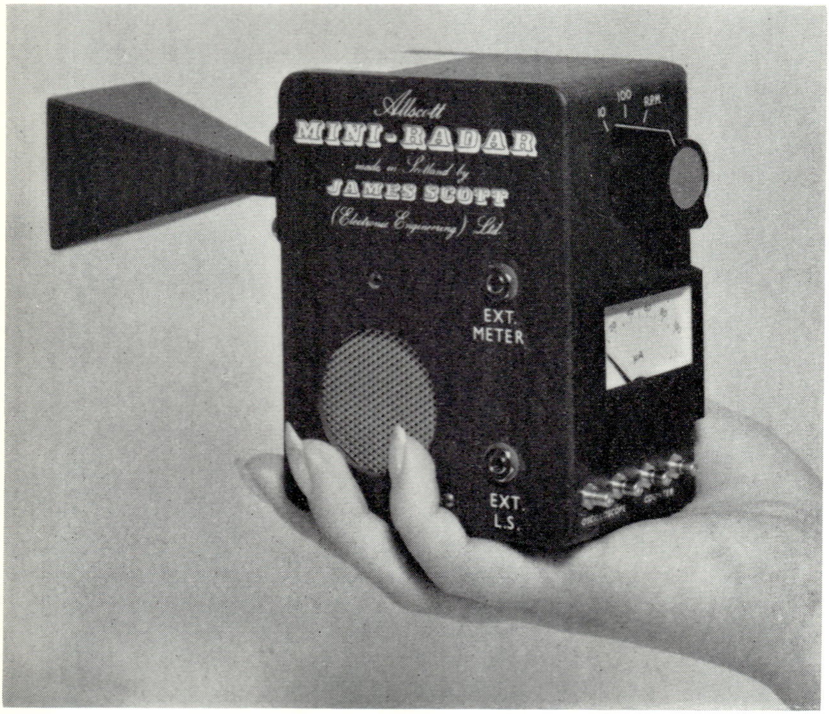

PLATE 15. A commercial speed radar, from which dynamic proximity signals can be extracted at ' ext L.S. ' jack socket. The device uses Gunn diodes.

Inevitably this heading overlaps the preceding Section 2—since all proximity probes are displacement transducers. Hence this section, in respect of linear displacement transduction, deals with those devices which are attached to or part of the mechanism under test. One of the greatest difficulties in using displacement transducers is in obtaining a true reference point. Therefore unless it is necessary to measure the relative movement between two closely adjacent parts, a preferable alternative is frequently to use a seismic or acceleration pickup and deduce therefrom the displacement—which will be that referred to the earth's reference frame and not any particular part of the engine (see § 7).

3.1. *Linear displacement transducers*

The commercially available types fall mainly into two categories. Those made for very accurate control of numerical machines often give a digital output directly and are rarely used for vibration measurements. Those with an analogue type of output, either a direct voltage change or modulation of a carrier are suitable for the field presently under discussion.

One very common type, consisting of one or more contact brushes working on a wirewound or moulded resistance track, have value for movements up to several centimetres, as long as the frequency being measured is really low, i.e. under, say, 20 Hz. However, wear is generally a major problem and the limited number of turns of the wirewound type rarely gives a resolution better than 0·5% of full scale. They are not often used for vibration measurements for the above reasons, but are often required, in conjunction with vibration pickups, to determine the setting of engine control parameters relevant to, or the cause of, engine vibration. For example, the position of inlet guide vanes, bleed valves or throttle position.

One outstanding advantage of these directly coupled resistive types is the almost complete absence of cross sensitivity—as long as the mechanical coupling is properly designed so as to ensure movement only in the desired axis of measurement. Also they can have a very large output, up to 10 V or more and are low impedance devices. Another common type consists basically of a rod coupled to one reference point carrying a slug of magnetic material moving in a housing carrying one or more coils. The transduction is performed by allowing the slug to alter the back e.m.f. of a single coil; or preferably differentially vary the inductance of two coils.

In this type the polarizing of the coils is generally by a power oscillator at from 3 kHz upwards, thus enabling the transducers to be used with the multichannel carrier equipment generally available for other purposes.

While this type too is almost free from the effect of cross-axis movement, as far as the accuracy of read-out goes at lower frequencies, it is not suitable for situations where high vibration at high frequencies along the cross axis is present, since wear of the guide bearings as well as fatigue of the operating rod is likely; while vibration of the slug in its clearance may introduce a modulation of the dynamic effects being measured in the other axis. For this reason the inductive type is rarely used on engine test, though it is useful for rig work where high resolution is required. Plate 16 shows a selection of displacement transducers.

3.2. *Angular displacement transducers*

For control purposes it is quite common to use a linear or circular resistive element coupled to the angular movement directly or by bell crank and lever, but this is generally quite unsatisfactory for dynamic work over 30 Hz. Apart from the nonlinearity which occurs in the linkage and a high wear rate, there is also the limitation to, say, a maximum of 100° of arc with some types.

Since the requirements of vibration measurements often imply the necessity of an extended frequency response, other methods have to be invoked. A synchro-transmitter, such as was originally designed for transmitting the readings of compass and other aircraft instruments to a distance, can be used as a dynamic device if its circuit is slightly rearranged and polarized with a suitable carrier voltage.

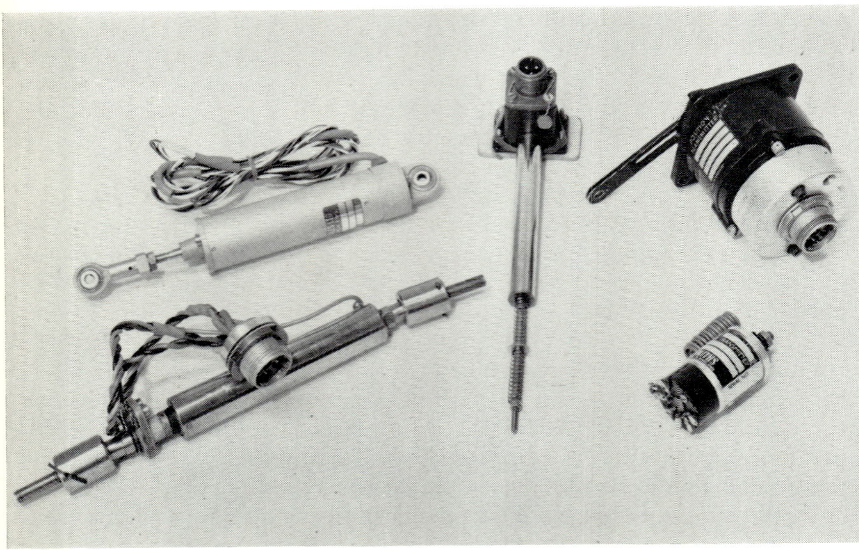

PLATE 16. A selection of resistive and inductive displacement transducers. The three on the left of the picture are linear displacement devices. The top right-hand one is capable of linear or angular transduction. The bottom right is a synchro transmitter which will transduce angular movement up to 180° of arc.

True it becomes ambiguous, used in this manner, over angles above 180° but it is reasonably linear over about 100°, and its dynamic response at the top end of the range is limited by the frequency of the carrier used. However, the latter can be as high as 10 kHz so that response to 1500 Hz can be secured—and up to 500 Hz even with 3 kHz carrier. The resolution is excellent, datum shifts are negligible; also because the devices are made in large numbers to tight Government specifications, the units are freely available in a number of standard frame sizes, and are easily interchangeable.

For such requirements as recording flutter in swirl vanes or rotary controls, where the movements do not exceed 100°, synchrotransmitters are very convenient—especially as they are compatible with carrier polarized systems used for pressure transducers and the like.

Where it is necessary to measure the angular displacement between the two ends of a shaft—as would be required for torsional resonance problems—perhaps combined with that of the steady torque, the shaft may be straingauged to form its own transducer. In this case if polarized with d.c. the response is limited only by the readout equipment. Carrier polarizing can also be used.

Although a separate section has not been thought justified for them, angular accelerometers can have their output double integrated to give dynamic angular displacement. Such devices generally comprise a flywheel ring, or disc, coupled to the part by straingauge spring arms, or with piezoelectric crystals interposed for the same purpose. There are not many angular accelerometers on the market and they cannot often be used in aero engine work because of the lack of room to put them where they are most often required.

The torsional problems on the larger shafts (compressor and turbine assemblies) would be so quickly lethal that they would hardly survive the earliest development tests; and it is easy to predict the resonances at the design stage and make sure they do not fall within the range of engine speed expected. Torsional resonance effects within gearboxes will require local straingauging—who ever saw a gearbox with any room to spare for anything but its own parts, often barely that?

Such cases, as were detailed in Chapter 5.2, where torsional resonances occur, may often be conveniently solved by having one or more toothed discs working with an electromagnetic probe; or similar discs with a large number of holes scanning past an electro-optical pickup. If the number of impulses is correctly chosen they generate a ‘ carrier ’ the frequency modulation of which, when extracted, provides the information desired. Indeed by using a high speed differential counter or high resolution phasemeter it is even possible also to determine the steady twist on which the dynamic fluctuations are superimposed. For these purposes the passing frequency of teeth or holes should be some decade higher than that of the dynamic flutter to be measured.

Some of the commercial torquemeters have facilities for taking out fluctuations in the torque for dynamic purposes.

4. *Pressure transducers*

As mentioned earlier several of the basic transduction elements are made use of in pressure transducers. The mechanical design of these transducers, as they are used for dynamic purposes, are all basically variants on the Bourdon tube or the diaphragm. Most of the time the deficiencies of such transducers are caused by mechanical drift and hysteresis rather than by the electrical readout part of the device.

Taking the earlier application of the Bourdon tube type a simple way of extracting the desired information is to make the deflection of the Bourdon tube move an arm over a rheostat. Such transducers are limited to about 30 Hz, are badly affected by extraneous vibration and wear rapidly when subject to appreciable dynamic variations on the steady pressure; their resonant frequencies are also low and their resolution is about $\frac{1}{2}\%$ full scale. They were of course not designed to measure dynamic effects.

Capsule operated transducers—suitable though they often are for low pressure control purposes—are inevitably large in internal volume, low in stiffness and consequently ill adapted to measurements of dynamic effects above about 30 Hz. Having thus dismissed the foregoing as usually unsuitable for dynamic measurements we are left with diaphragm devices as our preferred type for pressure fluctuations. Though the Dean gauge consisting of a short oval straingauge tube is still successfully used for many applications, it is rather inclined to fatigue failure of the tube if exposed to high fluctuating pressure changes.

Having obtained a satisfactory diaphragm design—and the diaphragm thickness may vary from 0·010 in. to 0·100 in.—the deflection can be sensed in several ways. Those devices discussed as proximity probes can all be used, or we can use straingauges of simple or special forms fixed directly on the diaphragm. Since capacitance probes are at their best over very small distances a diaphragm for this use can be very stiff and of a high resonant frequency. Inductive devices can also be made quite sensitive to small displacements and so they can be used almost equally well. Both the foregoing are adapted to carrier frequency operation or frequency modulation and do not lose the datum steady pressure, which is often required to be sensed as well as the dynamic variations.

Where the datum pressure information is not necessary, and particularly where very high frequency response, *circa* 20 kHz and higher is required, quartz piezoelectric pressure transducers may be used with advantage. In this type the pressure on the diaphragm loads one end face of a suitably cut quartz slice which develops a charge on the end faces. Charge amplifiers are required for accurate work and high insulation resistance leads are a *sine qua non*.

More recently silicon diaphragm transducers where straingauges are formed in or on the surface, also transistor type devices where a probe presses on a transistor, have been made. At this point in time these devices are still under development and evaluation; troubles due to hysteresis and drift appear to be more severe than they are with the older, fully developed, types. Some, at least, of the trouble appears associated with the clamping of the diaphragm in the transducer.

The writer, as of now, would use 3 kHz mutual inductance types, where not contra-indicated by the 500 Hz frequency limit, wherever possible. Above 500 Hz the capacitance type with polarizing around 2 MHz or the piezoelectric sort with a charge amplifier are usable, respectively, to 20 kHz or 50 kHz and even higher. A variety of pressure transducers suitable for dynamic work is shown in plate 17.

5. *Anemometers*

Anemometers, as used in general for recording wind speed, and airflow in ducts, are not useful for dynamic work—largely because of their slow response. The only types of value for rapid dynamic work are ' hot wire ' or ' thin film ' types. Both of these depend on the

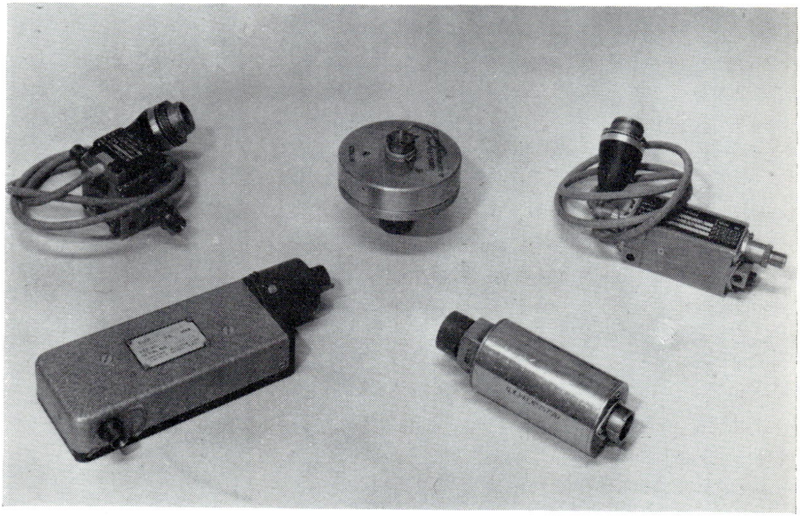

PLATE 17. A group of pressure transducers; bottom left is resistive potentiometer; top left is carrier inductive; centre top is a straingauged diaphragm type; top right is carrier inductive and bottom right is straingauged oval tube Dean gauge type.

same principle, i.e. the change in cooling produced on an electrically heated resistance wire or an evaporated film, usually of high resistance material. The element has to be maintained at all times at a higher temperature than that of the air—or gas—to be measured or, obviously, there would be no cooling. Plate 18 shows anemometers of three hot wire types, two free and one with the wire mounted on an insulated surface.

For accurate steady-state work rather elaborate precautions have to be taken in calibrating these devices, since their sensitivity depends on the running temperature of the resistance element relative to that of the gas. However, in the present context, where we are discussing dynamic changes in the gas velocity, such fine precautions are not usually necessary. The method of extracting the information from the change of resistance with change of cooling, can be either by having a constant current supply and sensing the change in potential difference across the element; or by using a constant voltage and sensing the change in current. In practice the former is more useful.

The thermal capacity of the element governs the maximum frequency of response, as with any other temperature sensor, so that fine wire or very thin evaporated films on a poorly conducting insulating substrate are needed for aero-engine work. The hot wire may, typically, be 0·001 in. thick tungsten 0·15 in. long and this will then give a response

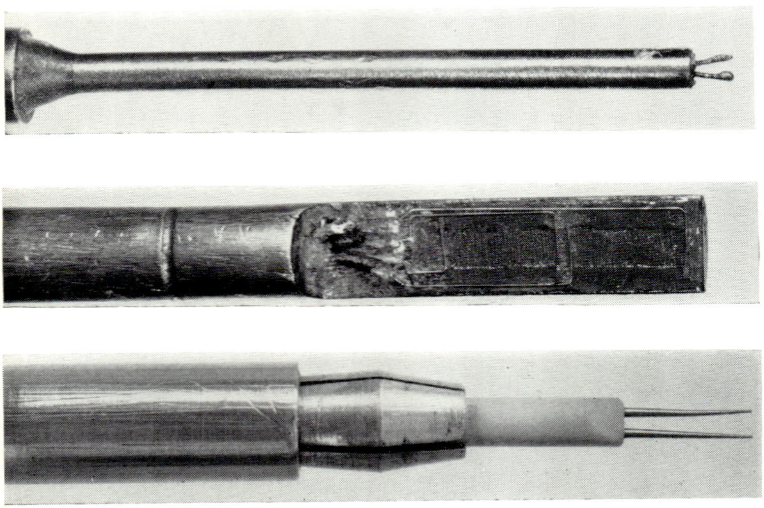

PLATE 18. Three anemometers, all essentially hot wire. The upper two have one thou tungsten wire for engine use, in the second one set in Brimor refractory cement. The lower one has 5 micron wire for use where clean air is available.

to high rates of flow—such as are found in a compressor—up to about 300 Hz. More sensitive types with 5μ wire used in research rigs where the air is clean can respond to 40 kHz.

Anemometers are most often used, in dynamic work, to measure fluctuations in velocity such as may affect a stationary part when the airflow is modulated by rotating blades. In addition a fruitful source of trouble, which results in rotor blade resonance, arises from the compressor mismatch at ' off design speeds ' causing rotating cell stall. This can be detected by a stationary anemometer. Finer wires improve the frequency of response but are more easily damaged in handling and by impact of dust in the engine operation.

Anemometers of this type are also used in noise research which is outside the scope of the present book. Since rapid velocity fluctuations must be accompanied by pressure changes, large cyclic ' velocity modulation ' will often be more easily read using a fairly sensitive pressure transducer connected by a (short) pipe into the region under investigation. Here the maximum frequency which can be read out is reduced, if too long a pipe is used, because of the greater volume of the pipe plus transducer. It is also desirable to keep the pipe length below that where the fundamental ' closed pipe ' longitudinal resonance approaches any of the frequencies to be measured.

It might sound attractive to use the type of anemometer with a vane on an arm—which is straingauged—and indeed for very low frequency effects in gas this is attractive—also for higher frequencies in liquid flow. However, it is difficult to raise the resonance of the vane type high enough to avoid fatigue trouble when measuring in gas flow, without undue loss of sensitivity and the vane constitutes a greater blockage than wire or film types thus modifying what is being measured.

Where frequencies of fluctuation in air are below 30 Hz and are of fair magnitude a more robust anemometer may be made using pitot-static tubes connected to either end of a differential pressure transducer, an inductive type of transducer is suitable for such application, with a carrier of, say, 3 kHz. The transducer can then be mounted external to the engine and a hole no bigger than 0·25 in. diameter can be used. The frequency limitations are normally determined by the volume of the transducer chamber and the length and bore of the tubes. The frequency response of this type can be improved by reducing tube and transducer volume; with special care it may be possible to raise top frequency by a decade.

6. *Microphone and stethoscope transducers*

These two types of readout are essentially similar. On the very simplest basis, it is possible to use a doctor's stethoscope on external surfaces of a running engine or on engine parts in rig testing. This particular type is peculiarly effective at very low frequencies—from 40 Hz downwards.

In using such methods one relies on the fact that the ear is a very effective ' wave analyser ' on line to the computer of the brain. Where the character of the effect being examined is changing it is amazing just how successful such direct methods can be.

When the internal parts of the engine require investigation, some less direct means is needed to protect the operator, and one is compelled to use microphones capable of withstanding the gas pressures involved, e.g. in the compressor. This can be done either by interposing a suitable thin diaphragm or by using special microphones capable of withstanding the pressures involved.

It would be unreasonable to attempt—in general—to put the microphone inside the engine so the sound is led out along a suitable tube. Think of the speaking tube used formerly in doctor's houses from front door to bedroom for night calls, and you will realize that legible speech can be heard over many feet of such tube.

Since such engine aerodynamic effects as rotating cell stall, and surge, cause pressure as well as velocity changes in the air it is often more convenient to use a microphone rather than an anemometer and the frequency response is higher under dirty air conditions. There are probe microphones available with tubes as small as 0·040 in. diameter.

Where extended frequency response (beyond about 1 kHz) is not needed, a very simple microphone which needs no polarizing supply, can be made from a standard G.P.O. telephone earpiece insert. Such a device is shown in plate 19. For determining such effects as rotating stall drop-in and drop-out no more is needed.

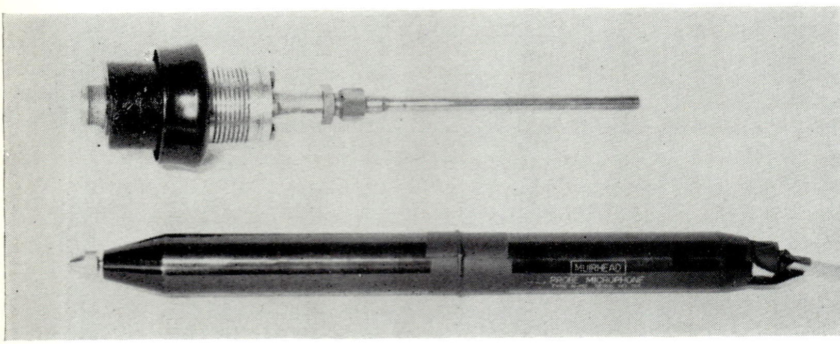

PLATE 19. At the top the stethoscope previously described, this will not respond to high frequencies, which is at times an asset. At the bottom a probe condenser microphone with extended frequency response.

For cases where blade passing orders, up to 20 kHz, are needed a condenser microphone is necessary and plate 19 shows one commercially available type. Questions of interpretation will be dealt with in Chapters 8 and 9.

7. *Vibration pickups—seismic and accelerometer*

Vibration pickups, be they of either description, depend essentially on one effect. This arises from the fact that basically they all comprise a mass mounted on a spring, as is explained in textbooks on dynamics. Such a system responds as displacement sensitive above the fundamental resonance (of mass on spring) and acceleration sensitive below this frequency.

Because the earliest displacement pickups were those used for measuring earth tremors this type is generally referred to as ‘ seismic ’ while the other type is an ‘ accelerometer ’. Although, as just stated, each type becomes sensitive to the other parameter, below or above the resonance, for practical reasons they are rarely used in reverse–in as far as their mechanical function is concerned. As mentioned earlier in the book, such pickups all give results relating their movement *to the earth* and not to any part of the engine or rig. The essential elements of a seismic pickup are illustrated diagrammatically in fig. 14.

Wholly mechanical vibration pickups are used for control purposes, for simple trouble shooting on engines in the field and for recording transient shock histories of packages and vehicles. They have no significant application otherwise, in the field covered by this book, and those types giving an electrical output are wellnigh universal.

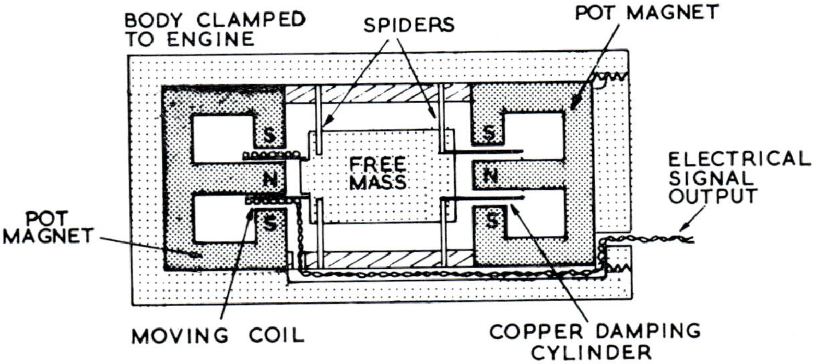

Fig. 14. Seismic moving coil pickup (diagrammatic sketch). Note: coil and damping cylinder are either much longer than magnet gap—or much shorter.

When using seismic pickups it is very easy (over a moderate frequency range) to differentiate electrically the output either once for velocity or twice for acceleration. Correspondingly when using accelerometers one can integrate once for velocity or twice for displacement. There are limitations, in the original output, in respect of frequency which will be discussed in Chapter 8, on such a process.

It is important to remember that pickups, of either type, read the movement relative to the earth and not to any other part of the engine; also that to obtain such relative movement it is necessary to combine the output of two seismic pickups or two accelerometers *vectorially* if the result is to have any sensible meaning; unless the frequency spectrum of each pickup is similar it is impossible to be quite sure what is happening between the two parts.

Considerations of space and weight, together with the range of frequencies likely to be found, may sometimes dictate the use of one type when, on the face of it, the choice appears incorrect. However, when this is done the differing frequency limitations of both types must be carefully considered.

Because the writer has found that many engineers have somewhat vague ideas of the effects of frequency on the relative magnitude of displacement and acceleration this seems a natural place to make some remarks. The mathematical expressions relating displacement and acceleration, of a simple harmonic motion, with frequency are well known; so well in fact that their implications are sometimes forgotten! For a constant peak alternating acceleration the displacement increases inversely with the square of the frequency. For a constant displacement the acceleration increases directly with the square of the frequency.

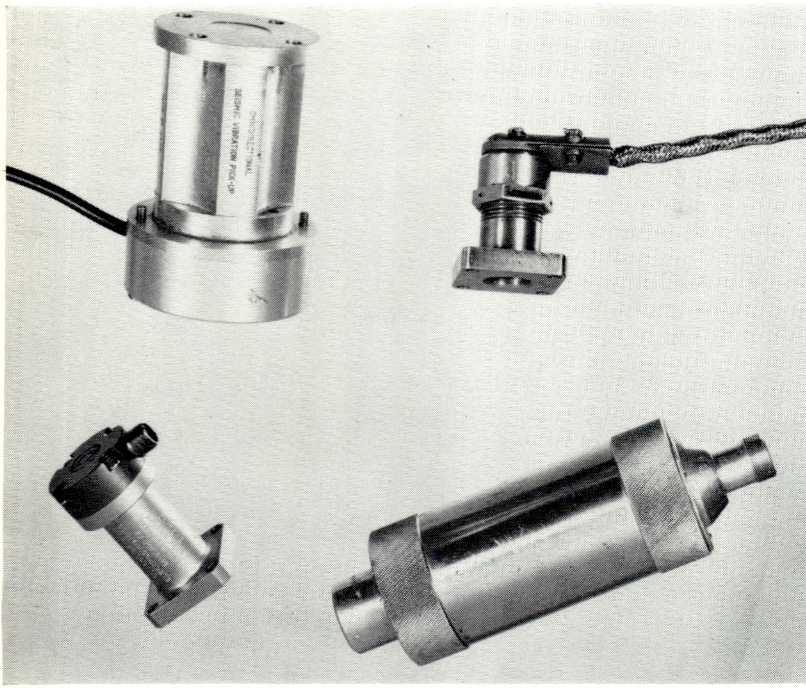

PLATE 20. A selection of moving-coil seismic vibration pickups.

Thus it is surprising that, quite often, experienced engineers ask for example for an accessory to be tested at, say 5 g from 3 kHz to 10 Hz! This seems innocent enough until the limit cases are worked out; now 5 g at 3 kHz implies a peak to peak displacement of about 10 *micro-inches*, while 5 g at 10 Hz results in a similar displacement of 1 *inch*.

In a typical case of a high speed engine, the above case would considerably underestimate the accelerations likely to be present in any gearbox at high frequencies. Equally the deflections implicit at 10 Hz would probably cause serious fouls between the engine and other parts of the structure; though when mounted in an aircraft wing the whole engine and nacelle might move by such amounts at still lower frequencies. All these remarks emphasize the need to avoid sweeping generalization and consider carefully what the requirements of each case demand. Plate 20 shows a selection of seismic pickups.

7.1. *Seismic vibration pickups*

Because a seismic pickup works above its resonant frequency this itself is generally low. For engine work it will be from 3 Hz to 25 Hz (for airframe work it would be a decade lower) hence soft springs and relatively large mass are required. It follows that the lower the resonance, the lower also the top frequency which will give adequate output. From this it will appear that seismic pickups are very much bigger and heavier than accelerometers. Typically they range from one inch to 6 in. long and from half a pound to 2 lb weight. Brackets for attachment are also somewhat massive.

The sensor method may be an arm moving along a resistance, not usually very satisfactory, a coil moving in a magnetic field; or a change of inductance of coils in a carrier-polarized circuit—often at 3 kHz. The varying resistance method and the mutual inductor both preserve the datum and can give steady as well as dynamic readings of this. The moving coil types suppress the datum level of steady acceleration, velocity and deflection.

Beside the above considerations, the resistance and mutual inductor types give an output linear with displacement at all relevant frequencies, but the moving coil type does not. While the deflection of the mass of the latter type is itself linear with imposed displacement the output from the moving coil depends on the *rate of cutting of lines of force* from the magnet. Hence the output of the moving coil type is velocity—not displacement—sensitive and requires a single integration to give displacement.

Moving coil pickups are often used at frequencies from 25 Hz to up to about 2 kHz ; mutual inductance for a range from 1 Hz to about 1 kHz, although as noted for acceleration, they can be used down to zero frequency. Several of the best designs of moving coil pickups dispense with spiders and guide the mass on a jewel spindle or in precious metal plated guides. While these avoid spider fatigue failures, they suffer from a small degree of ' stiction ' and may exhibit a threshold limit of, perhaps, 0·3 *g*—below which there is no output, this is particularly likely when the pickup axis is horizontal.

Almost all seismic pickups show appreciable cross sensitivity, i.e. an output arising from movement transverse to their sensitive axis. In addition, the moving coil types produce an undesired output from torsional oscillation of the body, which swivels the coil in the magnetic field. This generally occurs only near the top of the frequency range and is more often filtered out in the associated electrical readout.

The very limitation on top frequency response of a seismic pickup is sometimes an advantage since it constitutes a mechanical low pass filter and avoids overloading tape recorders or oscillographs with non-desired higher frequencies—*as long as these are not relevant* to the circumstances of the test.

Mutual inductor carrier-polarized pickups have particular advantages when it is desired to tape record the results, for later analysis; since the general limitations of direct tape recording impose a lower limit of no better than 25 Hz and often 40 Hz. Such a property is particularly convenient in random field testing when a domestic type tape recorder may be all that is available.

7.2. *Accelerometers*

Discounting the directly mechanical types, for the reasons given earlier in this section, almost all modern pickups used in vibration work are piezoelectric in function. They consist of a fairly small mass cemented to, or held in firm contact with, a disc or bar of quartz or other piezoelectric material. The latter is now often lead zirconate which has a much higher sensitivity than quartz for the same force.

These types of design have a desirably high resonant frequency, from 30 kHz up to as high as 100 kHz and are very compact. In fact, they can be made as light as 0·5 g and no more than 0·3 in. in any dimension. The sensitivity, naturally, is greater the larger the size, as the resonant frequency is lower. The principle of operation is shown in fig. 15.

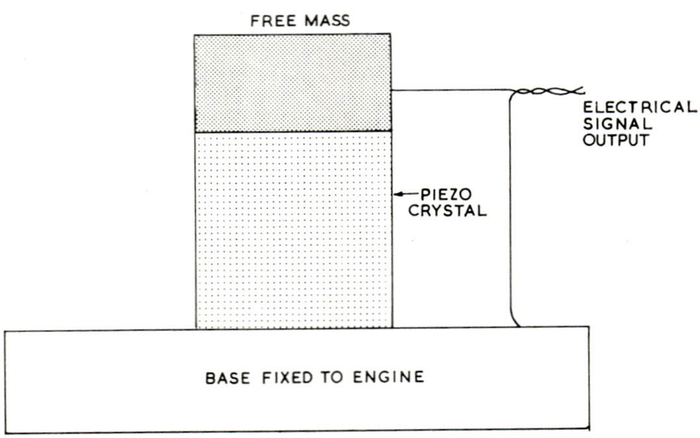

Fig. 15. Accelerometer (diagrammatic sketch).

In dealing with these devices the nature of the output must be kept in mind—a piezoelectric ' crystal ' produces an electric charge on its opposite faces when subject to strain (or force). Because of this the voltage output depends on the impedance of the following amplifier, an amplifier being always required because of the relatively low sensitivity. Hence solid-state preamplifiers must have an input impedance as high as possible. Also because the equivalent circuit of the accelerometer is a capacitance (in series with a small resistance) the capacitance of the leads reduces the signal with a conventional voltage amplifier, and the pickup must be calibrated with its lead wires, while the latter will have to be no more than, say, 6 ft long to avoid losing most of the signal.

A much better solution to this problem is now available in the ' charge amplifier '. This device is so circuited that it responds to the charge, or *quantity*, of electricity flowing through the circuit. Charge amplifiers may have leads, from the pickup, of almost any required length and their capacitance is, within broad limits, immaterial. However, the insulation of the leads must be of a high order. Charge amplifiers can be calibrated so that, if the charge sensitivity of the pickup is known, the result can be read directly from the output.

Because the accelerometer is so sensitive to very small displacements at high frequencies, the design of brackets to attach them to engine and rig parts requires special care. If attempts are made to use them, with the same brackets as are carrying seismic pickups, it is probable that resonances too high to affect the latter will vitiate the results from the accelerometer.

Generally the best results are obtained from accelerometers screwed directly on the engine part concerned, or if readings from three axes are needed, on a small solid cube similarly fixed. Alternatively a combined three axis accelerometer can be used.

The relative sensitivity of the crystal along the preferred axis depends on the perfection of its manufacture and therefore the cross sensitivity varies from crystal to crystal. Typical values range from 1% to 10%. If necessary, manufacturers may be prepared to select fine tolerances, however, the cost will be high as it is very difficult to measure this parameter and the test can only be done on the finished accelerometer. For normal cases 10% cross sensitivity does not matter, but where it is important to determine accurately the vector angles of a three dimensional vibration such steps may be necessary. Plate 21 shows some of the many types of accelerometer available.

The temperature limitations of lead zirconate transducers are presently about 250°C with the crystal the limiting factor. Development is, however, proceeding on pickups usable to much higher figures, now 400°C. Quartz does permit operation to 350°C and here the crystal is not so much the limiting factor, but rather the design of the mechanical part of the transducer and the means adopted to attach the active mass.

Where the accelerometers are merely attached to a hot engine part, then thermal barriers, of asbestos or other refractory insulation such as aluminium oxide, may be interposed between the base of the accelerometers and the job. In this case the shortcircuiting thermal effect of the securing screws must be minimized, and austenitic stainless steel or nickel chromium alloys should be used for fastening.

Where the whole transducer is in a hot environment the leads may well be the limiting factor. Mineral-insulated wiring is not very suitable as it is all too easy to restrain the very movement under measurement. In some cases the pickup and leads may be aircooled.

Because of the fact that the accelerometer behaves mainly as a condenser, changes in the capacitance of leads read as signal and the lead material should always be specified as anti-microphonic. For angular accelerometers see § 3 of this chapter.

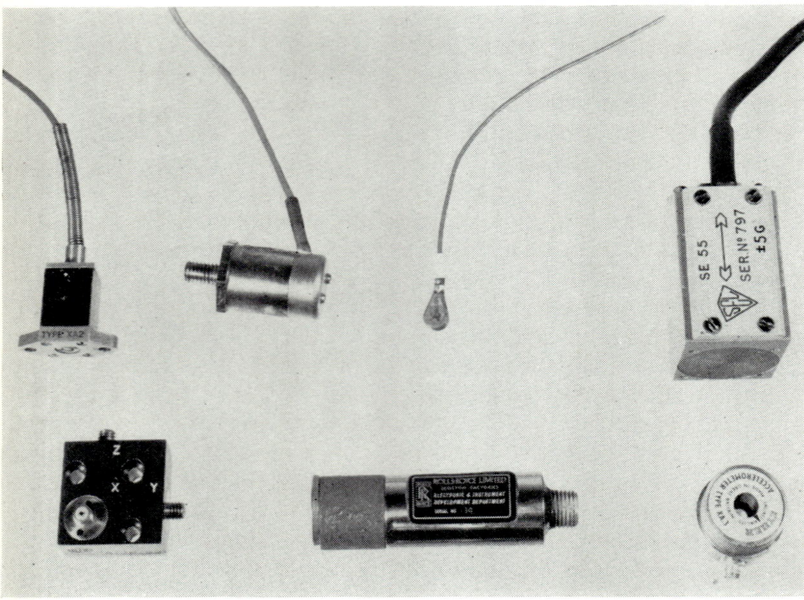

PLATE 21. A variety of accelerometers—all but the top right (which is carrier inductor) are peizoelectric types.

8. *Flow transducers*

Flow transducers are one of the rather more difficult fields of transducer design. First of all there is a very wide range of viscosity involved, which means a split between the low viscosity of gases and the (relatively) high ones of fluids; while in the latter field viscosities vary widely from water to heavy oils.

8.1. *Gases*

The measurement of velocity in gases has been dealt with under § 5 of this chapter. If the velocity is determined in a calibrated duct, then an anemometer becomes a flow transducer. However, when the duct becomes large it is not always easy to be sure of uniform velocity over the whole section.

Because of the difficulties involved—together with the problem of obtaining absolute calibration and consistency—it is often preferable to adopt the indirect method and measure the pressure drop across a section of duct or an orifice by suitable transducers as described in § 4. Again for vibration purposes, where rapid response is essential, great care must be taken that the system is not frequency limited either in the transducers or the readout equipment.

If large ducts are involved, then a number of pressure lines around the periphery with their inner ends projecting well through the boundary layer against the duct will be needed. If these are joined to a common manifold then the volume of the system will be large and the frequency response severely curtailed. Alternatively a number of separate transducers may have their electrical outputs averaged—not an easy problem. Differential pressure transducers would be used, of course.

If only the frequency of the gas vibration is required accurately, while the amplitude needed is only approximate, then a single anemometer or pressure probe or microphone will serve and the matter is fully dealt with in § 5.

8.2. *Liquids*

With liquid flows the forces on a transducer of vane type are much greater than with gas, and the resonant frequency of this type can be kept moderately high—say, 1 kHz with some sacrifice of sensitivity. On the other hand, because liquids transmit sound so well very high fluctuations of flow do occur, well up to 10 kHz or higher.

In these circumstances it is usual to infer high frequency fluctuations of flow from the corresponding pressure changes on differential pressure transducers, either end of a length of calibrated pipe or either side of an orifice; or in the case of flow through a burner or jet by single pressure measurements just upstream of the orifice—the burner having been previously calibrated by the weight or volume of fluid delivered in a known time.

The windmill type of flowmeter, often called a ' Pottermeter ' is a very accurate device for steady state measurements but is severely limited in rate of response for fluctuating flows. The speed of the rotor is limited to about 3000 r.p.m. by mechanical considerations, even when jewels are used, and any variations in the speed are in effect frequency modulation of the output signal. Many of the a.c. to d.c. converters have lowpass filters at from 15 Hz to 30 Hz built in, to avoid spurious breakthrough of higher, unwanted, frequencies. Therefore this type is unsuited for vibration measurements—except where these are below, say, 15 Hz. Nevertheless when investigating control system oscillations, which may well be between 3 Hz and 12 Hz, they can be useful. Plate 22 shows various sizes of this type of flowmeter.

The upshot of the above arguments is really that differential or single pressure measurements should be used—especially when the frequency spectrum of the fluctuations is unknown.

9. *Force and load cells*

While the force exerted on a known mass can always be deduced from the acceleration of that mass, this presupposes that the mass is rigid. Engine accessories and components rarely fulfil this condition and so, from time to time, it becomes necessary to transduce force directly. Also when investigating modulations on the steady thrust of jet engines, or rockets, the same need may arise.

PLATE 22. Three sizes of ' Pottermeter ' windmill type fluid flowmeters. These give a pulse output whose frequency is proportional to flow.

It is usual to call those force transducers used primarily to measure steady state force—or weight carried—' load cells ', while those used primarily for measurement of alternating effects are more often called ' force cells '. There is absolutely no difference in the function of either, though load cells commonly cover large magnitude forces up to millions of pounds force and force cells are more often made up to, perhaps, 3000 lbf.

The more usual construction of load cells is a straingauged tube, with flanges, carrying one or more full straingauge bridges. These are arranged to eliminate, as far as possible, temperature change error and also that due to asymmetric loading (with its concomitant bending moment). Force cells, as usually catalogued, are made with tubes or cylinders of piezoelectric material, preloaded to a high compressive strain on the crystal material so as to avoid the need for bonding under tension conditions (piezoelectric material is brittle under tension or bending). The construction is indicated in Fig. 16.

Load cells are normally used with 3 kHz polarizing and give readings on a dial or digital meter; however, when it is desired to measure the fluctuations in thrust the circuit may be tapped and cyclic effects up to about 500 Hz extracted. If higher frequencies must be measured the load cells may be d.c. polarized, when there is no practical limit to the frequency response.

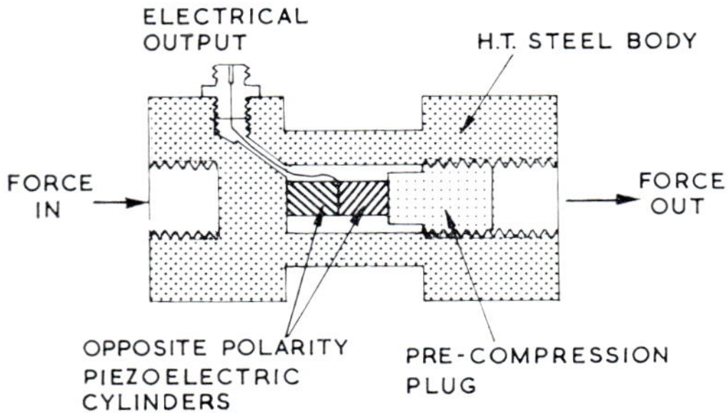

Fig. 16. Force cell (diagrammatic sketch).

Force cells of the piezoelectric type—usually lead zirconate, but they may be quartz—are used when most of the force acting between the two parts connected is of an alternating nature. Typically such a transducer will have a longitudinal resonant frequency from 25 kHz to 300 kHz—according to size and force rating. Two force cells are shown in plate 23, one of these being a straingauged type.

As noted, in connection with accelerometers, the same considerations indicate the use of a charge amplifier as initial readout equipment. Force cells are particularly suitable for resonance searches of components and accessories. They are essential for dynamic mechanical impedance and stiffness measurements, where they are used in conjunction with accelerometers, to read the acceleration resulting from the force at its point of application.

10. *Blade tip probes*

While these devices are really only proximity probes, used in a particular manner, they merit discussion in this application. Because of difficulties in applying telemetry or sliprings to certain rotating assemblies—notably the outer spools of multishaft jet engine compressors—there are times when an alternative means of detecting blade vibration is needed.

If a rotor blade is vibrating, as it spins round, then its angular position will not remain in constant relation to that of the rotor as a whole, it will as it were go out of pitch. If therefore a master timing pulse is taken each revolution, from the rotor, and the instant of passing of each blade is detected by a small coil or other probe at the inner periphery of the casing the impulses so generated can be made to reveal (on a time average basis) the amplitude of each blade tip. Plate 24 shows a suitable probe.

Because the method depends on the frequency of the blade tip being such that it passes a given probe at various states of phase (related to the rotor) each revolution, a single probe will not reveal an engine order excited vibration. This is because the blade would be at the same state each time if the vibration is synchronized to the rotor, and one could not distinguish a dimensional difference from this, apparently, steady shift.

If, however, the blades are vibrating at a frequency not integral with engine order, then successive revolutions will cause the blade to pass the probe at every state of its vibration. A television ' raster ' is set up on a cathode ray tube so that the frame frequency is that of the revolutions of the engine. If the output of the probe device is arranged to brighten the screen each time a blade passes, there will be a line of dots across the screen. If the blades are not vibrating, or if they are doing so at an integral order of engine revolutions, they will remain dots—though the line may not be straight, because of pitch difference. If, however, the blades are vibrating non-integrally, with respect to the engine, then each dot will be drawn out into a line (by the conjunction of dots at different times) and this line will show the amplitude of movement of each blade of the stage.

As noted before, it is a pity that this method does not give the frequency of vibration of each blade, at the same time there are many occasions—particularly when surge problems are being dealt with—when this is not of prime importance. The alterations required to the engine are minimal; only two probes being needed. One to take the impulse from the rotor for reference and one so situated that the blade tips of the stage being examined pass it by. It is thus possible to apply the technique without the more major changes necessitated by fitting sliprings or telemetry.

A much more sophisticated extension of the same principle was developed, some years ago, by the Bristol Aircraft Company (now the Bristol Engine Division of Rolls-Royce). It is colloquially known as the " Belly band system ". In this method the probes are replaced by a stationary zig-zag conductor running peripherally round the inside of the compressor casing over the blade tips of the stage. This conductor has many axial bars which are interacted by a small magnet in the tip of the blade being measured. As the blade passes each bar of the conductor matrix a pulse is generated and these pulses are frequency modulated by vibrations of the blade. From the resulting signal it is possible electronically to recover the frequency and amplitude of the blade tip.

If the blade is firmly held in the disc, i.e. if the root is firtree or dovetail in form, the stress caused by a given tip deflection can be related

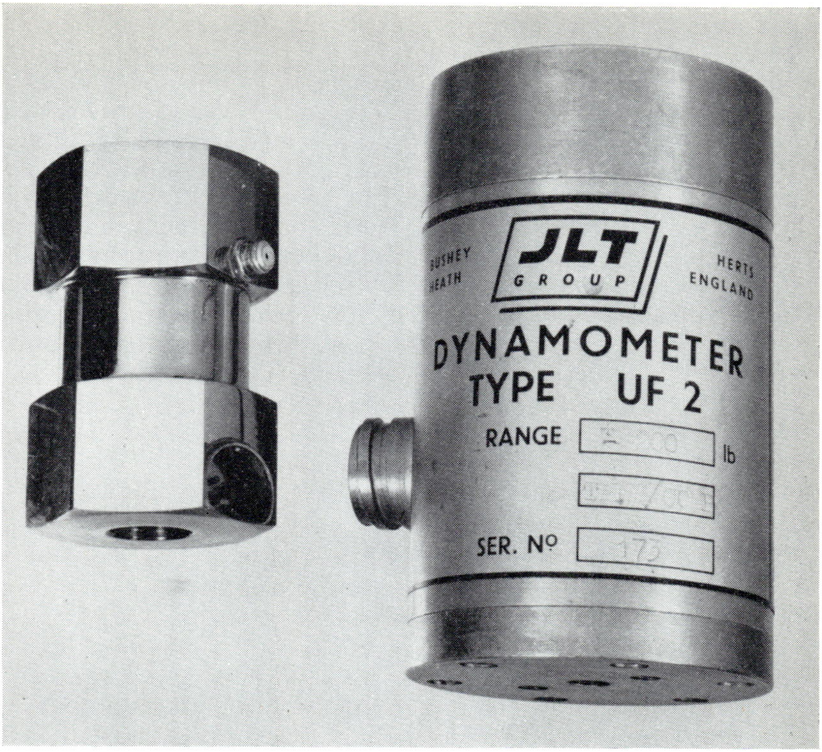

PLATE 23. A force cell on the left and a load cell on the right, both fulfil the same function. The force cell is piezoelectric while the load cell is straingauged.

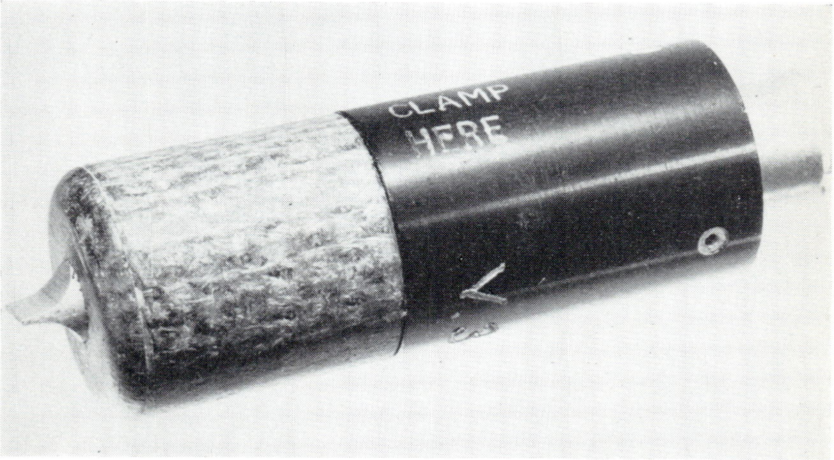

PLATE 24. An example of a self-generating electromagnetic probe, such as can be used to register the time of passage of blade tips.

to the stress by sound room tests on straingauged blades. For pin-fixed blades the system is less useful, since the stress can hardly be related to tip movement, though the engine condition which is worst can be found. Only one blade per stage can be tested at any one time and the calibration involves rotating the disc in some spinning rig.

In the days before telemetry and mercury slip rings had been developed to their present stage the system did much useful work. It is still used today, when circumstances make other methods difficult or impracticable, since it demands no addition to the rotating parts other than the minute magnet in the tip of the blade. It is however complicated to build and suffers somewhat from casing distortion changing the shape (and hence the pitch) of the conductor matrix, while blade tip clearance is also critical.

CHAPTER 8

application of transducers and recording of signals

THERE is a deliberate purpose in the duality of the chapter title. It is impossible to divorce the selection of transducers from the nature of the recording processes with which they are to be used—or indeed from the type of problem being investigated. This chapter, therefore, attempts to deal with the basic considerations involved in mounting a successful test, to yield the maximum of relevant information with the minimum of error.

1. *General precautions in selection*

Many of the problems involved in selecting the best transducer for a given job have been referred to incidentally in the earlier chapters. The present section is directed more specifically to the combination of several measurement parameters, and to the problems which arise in planning a complex test. Some comments will be made on the human interface which is inevitable between the engineer who has the mechanical problem, and the instrumentation engineer making the measurements.

Apropos the last paragraph the first question to be answered by the ' customer,' i.e. the design or development engineer, is what is really the problem. Such a statement sounds trite, but long experience indicates that more inconclusive tests—and even misleading results—arise from lack of clear thinking and agreement on this point, than for any other reason.

The difficulties often arise from communication difficulties between the two sides—the main reason for this book being written at all; at other times they spring from preconceived ideas on the part of the engineer faced with the problem. Sometimes, to be fair, they arise because the instrumentation engineer is not familiar enough with the engine problems. The writer cannot urge too strongly the importance of a full, frank, discussion of possible alternative methods of measurement, at the earliest stage in the planning of engine or rig tests.

Probably the most troublesome human cases arise when the customer has a little knowledge of instrumentation—oft times gained from the more popular scientific press. He may then well feel that he can dictate the type of measurement to the instrument engineer, without condescending to discuss the problem. If the latter is weak or junior to the former he may agree. Both may well discover, after expensive and time consuming tests have been made, that incidental effects have vitiated the readings taken.

Some examples of the above difficulties have been given, in the earlier chapters of the book, at various places. A few typical examples may serve to indicate the sort of trouble which may arise from lack of understanding and discussion.

1.1. The ' customer ' may ask for pressure fluctuations to be measured at a particular tapping point on the engine, not realizing that the same tapping will be connected to a manometer line, some 30 ft or more in length; in order to make accurate steady-state readings of the pressure in another part of the test. If the measurement engineer is unfamiliar with test bed or rig setup he may also fail to appreciate this possibility. Significant fluctuations in the pressure may well be missed entirely in such a case.

1.2. A development engineer may ask for straingauges to be fitted to components, particularly bearings, at the back end of a compressor and quote the temperature involved as that applying when the engine is running. This temperature may well be greatly exceeded when the engine is stopped, due to soak back from the combustion area, so that after the trial run up to set engine controls all the gauges are useless. Equally the customer may fail to remember that a straingauged component, normally quite cool, may be subject to anti-icing heating at some stage of the test; or it may be that this requirement is added later in the testing—too late to regauge the components.

1.3. Another type of case may arise when the occurrence of fatigue cracking is first assumed due to alternating stress alone, and the engine man fails to consider the possibility of a major transient differential thermal strain. The type of straingauging, and the manner of recording adopted, can be compatible with determining strains of both types, but this requires special gauging and circuits and will not normally be provided by the instrument engineer unless he is so advised at the beginning.

1.4. Another example concerns electrical interference from the component under test. The design engineer may want vibration measurements on an electrical generator attached to the engine and these may need to be made while the generator is electrically loaded. Unless he makes this clear to the Instrument Department at the outset results may well be abortive, since interference via the vibration pickup leads, even though these are normally screened, may completely swamp all vibration information. Very elaborate precautions, both in selecting the type of pickup, and in the recording equipment used, are necessary in such a case.

In selecting a particular type of transducer, for any specific application, one of the most important considerations is its sensitivity to other parameters than its primary function.

There is a good general rule: all other things being equal choose that transducer least sensitive to side effects, rather than that most sensitive to the main effect. An exception has to be made where unavoidable sources of electrical noise (such as high speed slip rings) impose an overriding demand for the maximum possible output.

Further restrictions on the choice of transducer may be imposed by the recording equipment available. For example, having settled the most important transducers in the best manner, there may be insufficient channels of, say, carrier amplifiers to permit using such methods for the remaining less vital parameters. Again, if it is necessary to tape record on machines equipped for direct recording only—as at some outstation or airfield, carrier-polarized transducers may have to be used even if less suitable in other ways; such a course would be necessary if standing pressure or effects occurring at less than, say, 30 Hz are important.

2. *Connection of transducers to recorders*

Various references have been made in earlier parts of the book to the way in which various devices work and their limitations, so repetition here is needless—in respect of their attachment to the engine. Difficulties do, however, arise when the connections of the various transducers of a complex test are involved. Crosstalk, i.e. the electrically induced signals from one channel to another of the recorder, can be very troublesome.

In the case of frequencies below some 250 Hz such pickup is mainly inductive—due to the magnetic fields, produced by the currents flowing in leads creating fields which penetrate the screening of the wires and induce similar, though smaller, currents in the leads of other channels. In such cases physical separation of the wires becomes necessary. Since the magnetic field is proportional to the current, these effects are generally restricted to low impedance circuits.

One might observe here that mains interference often comes into this category and is a constant source of trouble. There are appreciable fields, due to the electrical wiring, in most buildings at the fundamental frequency of 50 Hz in the United Kingdom (60 Hz in America) and at its lower overtones. The effect of local iron saturation in sub-station transformers is to create significant impurity in the wave form of the supply, mainly at the odd overtones; i.e. 150, 250, etc., the lower of which can appear as crosstalk due to magnetic induction and the higher as capacitative transfer.

While still dealing with the lower frequencies, and especially 50 Hz itself, an effect is due to what—in hi-fi or electrical terms—is described as ' earth loops '. If part of the measuring circuit *or its screening* is common to interference currents and signals from transducers then the latter may contain so much at 50 Hz as to swamp the genuine signal. Even if this is not quite the case the insertion, into the signal, of even

10% of mains interference will make visual monitoring tiring and subsequent processing of data difficult.

The best cure for earth loops is to prevent their occurrence! This means that, since there will always be an electrical path of sorts between the engine (or rig) and the test apparatus, the latter should have no direct connection to the engine. This is secured by connecting all screening to the test apparatus and not to the engine. Screening is bonded thoroughly from readout equipment to that of the last bit of cable going to the transducer, but is there left unconnected.

Unfortunately some transducers have one side of their output connected to their metalwork, in such cases it may be possible to insulate them electrically from the engine or to interpose a screened isolating transformer. If this is not done it must be remembered that not only the transducer involved, but all the others may be affected, if the various channels of the test equipment are interconnected at their ' earthy ' side.

Since earth loops can be equally harmful between various units of the (often elaborate) recording equipment, it is generally desirable that all of these should be centrally earthed at a common point *and not from one via another and so on.* There are, of course, local exceptions to the above remarks, but if adhered to in general much unnecessary trouble shooting will be avoided.

Capacitative crosstalk occurs preferentially at higher frequencies and its effect is the worse the higher these are. In general, screened leads will prevent this effect being of consequence, but at frequencies above 5 kHz quite short lengths—only perhaps a few inches—of wire will give trouble if not screened. The capacities existing between contracts on gauge selector switches and the like are a frequent source of trouble, as is the internal wiring of many multichannel recorders and amplifiers.

Since capacity transfer depends on voltage it is at its worst in high impedance circuits, as in piezoelectric devices, or straingauge signals fed to a high impedance amplifier. Also when, for example, a straingauge goes open circuit quite high levels of interference can take place, and be transferred thus to the other channels of measurement.

A less obvious source of high frequency interference occurs in parts of the engine where combustion gases circulate past components with straingauges. Appreciable ionization occurs in the products from hot flames, and these can charge the component if it is not firmly earthed and appear in the output from straingauges (capacitatively coupled to the part) as noise voltage interference. In this case the recommendation to sever paths between engine and test gear may have to be flouted and a wire run from the engine part to the equipment.

In the case of straingauges on rotating parts of turbines this trouble is often severe and one slipring channel may have to be used for the connection—in this context often broadly described as a ' Rotor Earth ', which it is solely in respect of the test equipment.

3. *Processing output before recording*

An important consideration in recording—either on magnetic tape or directly from oscillograph or galvanometer—arises from discrepancies in time scales.

The transducers used for various parameters may well give outputs at very different frequencies, sometimes several orders apart. For example, a typical test might record engine speed, compressor pressure fluctuations, rotor blade straingauges and two vibration pickups. If selection of transducers is made chiefly on availability, the tachometer standard to the engine would be used for speed, a straingauge or inductance type pressure transducer for compressor pressure and seismic pickups for carcass vibration.

Suppose then that the straingauges on the blades are giving signals at up to 3 kHz, while the engine speed signal (near half engine shaft speed) is around 70 Hz, and pressure fluctuations on the compressor show about 7 Hz, we clearly have an analysis problem! If every parameter is recorded as it comes out of the normal readout gear we shall have signals at frequencies ranging from 7 Hz to 3 kHz.

If graphical recording is being used, such a speed as will enable us to resolve separate cycles at 3 kHz may be necessary. A typical speed for the latter would be 2000 mm per sec, at this speed a shaft speed signal, say at 6000 r.p.m., will be pitched at 40 mm (tachometer at half shaft speed), while the 7 Hz pressure pulsation will have a pitch of about 289 mm! If the lower frequencies are sinusoidal it will be wellnigh impossible to define their maximum positions on the recording paper with any pretence to accuracy.

One of the ways of dealing with this dilemma is to ' shoot ' all the graphical recordings at a lower speed, say 200 mm per sec, which will give only an envelope of the blade stress amplitude, and then do short shots at 2000 mm per sec to define the frequency of the 3 kHz blade signal. A better method is to tape record at a speed high enough to resolve the 3 kHz frequency—but then f.m. recording will be necessary to preserve the 7 Hz signal. On replay differing speeds can then be used at will. Figure 17 shows low and high speed shots of the same engine effects.

If the frequency, but not the amplitude, of the 7 Hz and 70 Hz signals is all that is required (and this is certainly the case with the speed signal) then both these signals can be pulse-shaped (peak sharpened) by suitable curcuits so that they appear as ' blips '. If this is done it will also be possible to use direct tape recording as long as the equivalent frequency of the blip is above the lower cut-off of the tape recorder.

Another solution to this type of problem is to use frequency multiplying circuits for the lower frequency parameters, again only when their timing and not their amplitude is what matters.

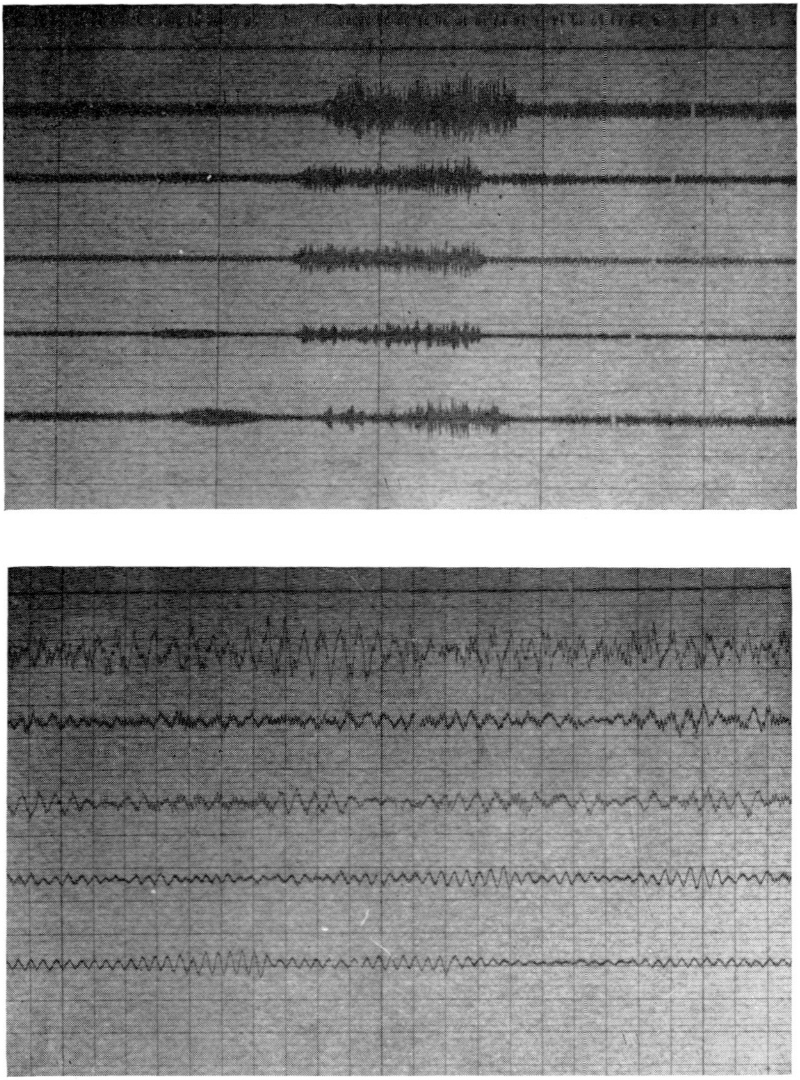

Fig. 17. Two shots of the same engine recording, one at 5 cm per sec and one at 100 cm per sec. The high speed shot enables the frequency composition to be seen, while that at low speed shows the stress related to the engine manœuvre.

The above epitomize the types of problems facing the vibration engineer, and the examples chosen are not by any means extreme! Perhaps it might make the customer think more kindly, when he wants to know why the instrumentation department is so fussy about sufficient time being allowed to gather together the equipment; and to plan the testing as regards the recording, as well as the engine considerations.

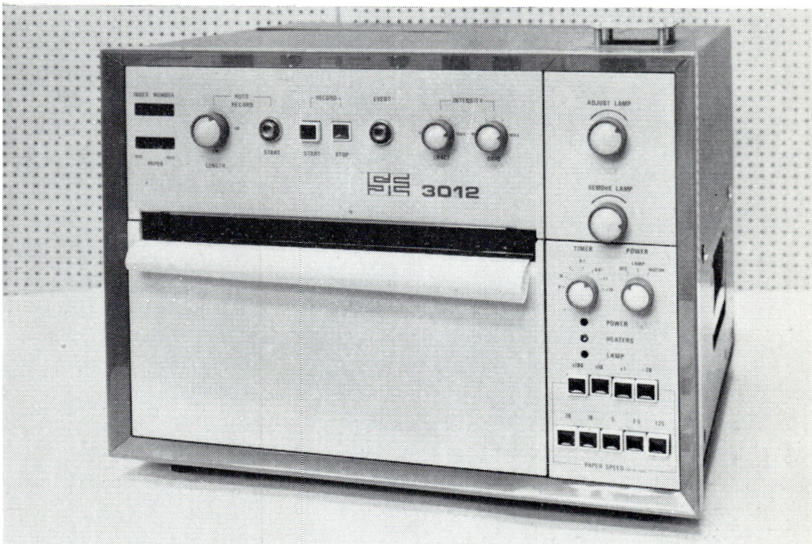

PLATE 25. An example of a u.v. galvanometer recorder of 50-channel capacity and capable of running at up to 125 cm sec.

Another case where it is essential to make adequate preparations, prior to the actual testing, occurs when the relative phasing of certain engine effects must be preserved in the recording. Suppose vibration pickups are sited along the length of the engine, in order to find not only the amount and frequency of the carcass vibrations, but also whether the engine is ' rocking ' or translating bodily. In this case the transducers must be of identical type, they must all be connected so that ' up ' and ' down ' give correct polarity of signal, and the recording channels must be checked for phase delay over the relevant frequency band.

A typical example of the need for forethought is when pressure pulsations along the length of the engine must be correctly phased; in this case not only must the transducers be of identical type, but they must communicate with the appropriate parts of the internal airways by tubes of equal length and diameter. Again the recording channels must be checked for phase and the transducers for correct polarity. Plate 25 shows a modern example of a 50-channel u.v. recorder.

A dilemma which repeatedly haunts the vibration engineer is whether to record all the information from each transducer over the full frequency range, or to use electrical filters to suppress frequencies in which he is not interested. The snag here is to know those which are not wanted.

If the full range frequency output of all transducers is recorded, that level which avoids ' out-of-scale ' direct recordings will clearly make it very difficult to detect small amounts of information which might, perhaps, be crucial. For example, if a compressor rotor blade strain-gauge signal is being dealt with, *and it is known that a first flap failure is being investigated,* then electrical filtering with a low pass filter with a roll-off at twice the expected frequency will remove most of the slipring and other frequency noise. The resulting records will be far easier to analyse, but imagine the red faces if later it is found that the failure might have been due to some higher blade mode, which has been so carefully removed from the results!

One way in which part, at least, of such trouble may be avoided is to record everything—straight and unfiltered—on magnetic tape, and filter the graphical recordings. Even here there are snags, as the usable dynamic range of tape recordings is rarely more than 200/1, and much the same ratio holds for graphical analysis—unless filtering is employed. If, therefore, there is a large amount of ' noise ' in the signal it will be necessary to use selective data reduction methods, many of which militate against accurate instantaneous readings, as explained in the next chapter.

Electrical filtering for frequency may be done in several ways, bearing in mind what has already been said about its dangers. Firstly it is quite simple to cut off frequencies below a certain value, such circuits are called ' low pass ' filters; or those frequencies above a particular value with ' high pass ' filters. Also if these are combined into a ' band pass ' filter one can admit—for recording only—a particular band of frequencies. This band can be almost as narrow as desired—even only a few cycles wide—but in this connection it is important to note that if too narrow a filter is used results will be unsatisfactory.

An engine running (nominally) at a particular speed, under control of some type of governor, speed, pressure or temperature sensitive, will necessarily ' hunt ' or ' flutter ' about a mean speed, as one or other limit of control is approached. Hence, even for a machine at only one speed, a highly selective filter may partly remove signals which arise from an engine order cause. If, as more generally happens, tests are made over a speed range of, say, 3000 r.p.m. to 9000 r.p.m., a filter to pass any individual engine order will need to pass signals with a frequency range of 3/1, i.e. an octave and a half.

There are tracking filters now available which will take an engine order signal (e.g. from tachometer) and electrically alter the centre frequency of its pass band as engine speed alters. More often these are used in replay than in recording, so they will be described in Chapter 9.

If the vibrations of the whole, or a part, of an engine are to be recorded and there is, for example, only room for an accelerometer; then an electrical single or double integrating circuit will give a velocity or displacement conscious output. This will enable more prominence to be given to lower frequencies than would be the case with an unmodified record.

Such circuits, however, may only work satisfactorily over a limited range of frequencies, generally about 100/1. Within this range will have to be included, not only the fundamental, but any significant overtones in the signal.

Correspondingly, the output of seismic moving-coil pickups can be single differentiated to record acceleration; or single integrated to give displacement. Remember that, though the mechanical part of moving coil seismic pickups is displacement sensitive, the output from the moving coil is velocity responsive—see previous chapter. The output of carrier-polarized induction type seismic pickups is displacement sensitive and needs single or double differentiation to give velocity or acceleration respectively.

When recording such parameters as the output of tachometers, or spinning vane type flowmeters; for graphical analysis at low recording speeds, it is quite common to convert the varying frequency to a d.c. analogue. This gives a sloping line instead of a multicycle trace on the recording and saves much analysis time—at the expense of some accuracy.

Spectrum analysis is another type of processing which can be applied, with modern electronic techniques, to signals as they arise from a running engine; however, at the time of writing it is more often left to be used on tape recorded data and will be covered in Chapter 9.

The final technique which will be mentioned, in this chapter, is X/Y plotting. Again this can be used from tape at a later date. However, it is one which is frequently used in real time on live tests. In most of the work on vibration all other parameters are recorded against a basis of time. In many cases it is because this is the most significant plot, in others it is used because it is easy to move a recording film, or paper or magnetic tape at a constant speed—which at once settles the matter.

If, though, instead of taking the direct output of the vibration transducer, this is passed through suitable circuits to produce a d.c. analogue of the cyclic amplitude and the same is done with, say, the engine speed signal; these two quantities can be crossplotted on an 'X/Y plotter'. This will give a graph which reveals more clearly the interrelation of the two parameters than the normal records against time.

However, the method is not suitable where the quantities being cross-plotted change at high rates; since the systems for moving the cross-plotter stylus have much greater inertia than the cathode ray beams, or the recording galvanometers, normally used. The full scale response of a modern cross plotter may require several seconds if attenuation is not to be introduced. Further, if significant attenuation does occur, the amount will vary with the rate of change of the affected parameter which makes it difficult to sort out the errors. Plate 26 shows a commercial X/Y plotter.

4. *Telemetry and slip rings*

A section on this subject seems necessary, since the recording of signals from transducers situated on moving parts is not always a simple matter. The most frequent reason for the use of such means in the signal channel is when the transducers are necessarily situated on a rotating component. However, there are also many cases of reciprocating parts from which signals have to be obtained, e.g. diesel engine pistons and connecting rods.

The man in the street, used as he is to ‘ walkie-talkie ’ devices as used by police and service authorities, sees no difficulty in putting a small transmitter on the moving part and a receiver somewhere handy—nothing could be further from the truth! One of the notable characteristics of radio waves is that they do not penetrate metal! In the majority of cases of vibration met with in engines and other machines the moving parts are surrounded by the casing of the machine.

The first difficulty then, in using telemetry, is of siting the receiver inside the casing of the engine, or at least so positioning the aerial system. The insides of engines are usually rather cramped and often filled with oil mist, hot gases, etc. Furthermore, there are numerous metal parts moving about which would obstruct or distort the path from transmitter to receiver.

4.1. *Rotating parts—telemetry*

For the above reasons it is generally necessary to make the transmitter and receiver aerials in the form of inductive coils, with the rotors and stators in fairly close proximity, so that the movements of nearby parts does not significantly alter the transmitted signal. This is especially important when more than one channel of transmission is involved, otherwise interaction between channels will make accurate measurements impossible. Plate 27 shows the type of coils needed in order to obtain satisfactory results.

Considerations of space, and the inevitable close proximity of metal parts, leads to the need to use frequencies for the transmission, between, say, 30 kHz and 500 kHz.

Because changes in the transmission path, due to variations in attenuation of the signal, are likely to be appreciable, f.m. modulation is highly desirable. However, f.m. modulation presupposes an oscillator (whose frequency can be modulated) adjacent to the transducer, i.e. on

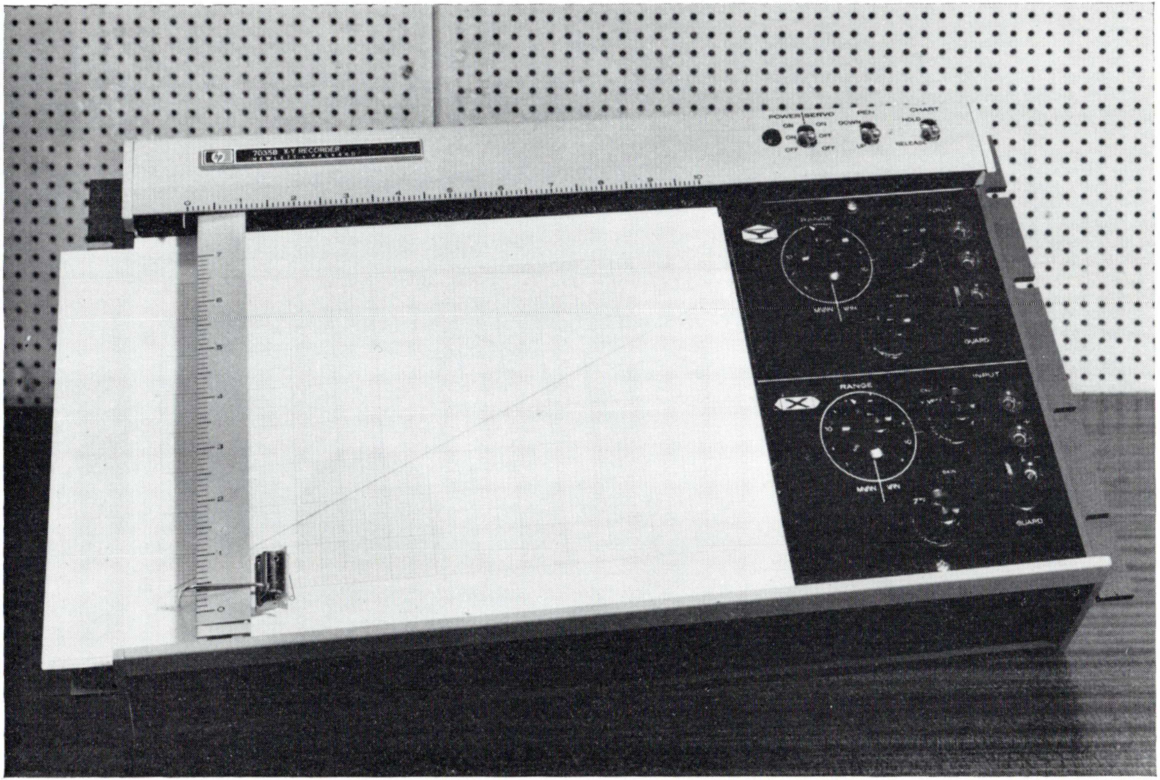

PLATE 26. One make of X/Y recorder, used for making cross plots of related parameters—such as fuel flow with engine speed. The trace shown is a calibration.

the rotating part. In that location the electronics are subject to high acceleration (due to centrifugal force) and probably high temperature.

Until a few years ago solid state electronics were limited to about 100°C, whereas now the latest devices will operate to 250°C. This has made f.m. rotating electronics a practicable proposition; as long as the whole pack is potted, to withstand the high acceleration, with resins capable of withstanding the temperatures involved. The problems are difficult but not insuperable.

PLATE 27. The inner stator coil assembly of radio frequency slip rings, using amplitude modulation. There are eight power input coils and eight signal output coils. The size is necessary to clear the shaft of the low pressure compressor which runs inside.

Earlier systems had to rely on amplitude modulation and were much more prone to calibration shift; and to modulation of the desired signals by engine order owing to eccentricities of the coil systems.

Even the newer systems still need power supplies; for cold rig jobs these could be batteries rotating with the job, but the power demands of straingauges, for adequate sensitivity, would mean a strip at frequent intervals to change batteries. In the present state of the art, rotating f.m. electronic modules are generally supplied by a pair of rotor and stator coils working at a lower frequency than any of the channels—say, 10 to 20 kHz, which is rectified and stabilized on the rotor to supply the FM modules and transducers. Plate 28 shows a module for f.m. as would be carried on the rotor being tested.

4.2. *Rotating parts—slip rings*

Because of the cost and complexity of the telemetry systems, together with the difficulty of engineering enough separate simultaneous channels, slip rings are still used whenever possible. Slip rings with up to 10 simultaneous channels are common enough. Until a few years ago they were often made by the user firms, because of the difficulty of obtaining a noise level appreciably lower than the output of the gauges. Nowadays, however, several firms make very good multichannel slip rings usable up to 10 000 r.p.m. (or even higher), with a noise level adequately low for straingauge use.

The type pioneered by Mr. D. A. Drew of Rolls-Royce have given satisfaction over many years, at speeds up to 20 000 r.p.m. and with a noise level below 100 μV at 10 mA current. These consist essentially of a capilliary film of mercury a few thousandths of an inch thick, held between amalgamated copper plated surfaces. While eminently satisfactory they are very costly to make and require frequent overhauling. Even today they are hardly surpassed by commercially available slip rings in compactness and noise level. Connections are made to the rotor of the sliprings, through the driving coupling, which latter has to allow for any relative axial movement of the rotor with respect to the casing. Filtered cooling air (or in some designs demineralized water) must be supplied to the slipring assembly during the test. Special checks are necessary if the output of thermocouples is taken through the sliprings; as in some cases local heating can produce spurious electrical potentials in the sliprings, commensurate with the output of the thermocouples. Plate 29 shows Rolls-Royce mercury slip rings.

4.3. *Reciprocating parts*

While slip rings can be used their sliding contacts must be linear (not circular) for obtaining the output. Because of the impulsive nature of the relative motion such devices have not been very satisfactory in the past. In such cases, and particularly when an easy line of sight exists between transmitter and receiver, telemetry can be used. Higher frequencies and smaller aerials can be used than with rotor and stator coil coupling. Such designs tend to be made as ‘ one off ’ devices tailored to the particular engine and location. For measuring cyclic temperature effects in diesel engine pistons and similar jobs, they can be very useful. Temperature compensation is not easy but can be satisfactorily attained. The author has had little experience with this type of problem.

5. *Recording data from transducers*

When it comes to conducting an actual test there are several options, which may be used separately or conjointly. In the simplest case of

PLATE 28. This shows an f.m. solid state module suitable for use on the rotor of a jet engine, before and after encapsulation in epoxy resin. It is the size of an ordinary matchbox.

all, particularly when qualitative results only are needed, it may be enough to display the output of one or two transducers on a cathode ray oscilloscope; either with multiple beams or with an electronic switching device and a single gun tube. In almost all cases the relation of two quantities is the minimum requirement. Polaroid photographs can be taken of parts of the test.

PLATE 29. Three Rolls-Royce ten-way mercury slip rings. The one at the top right was specially modified so as to build it into the coaxial shaft arrangement of an epicyclic gearbox.

Since most customers require proof, a graphical record, or a tape recording which can be replayed on to a graphical record is generally better. Again since tests are costly it is generally required to record from 5 to 50 separate transducer outputs. As noted in the previous section it may be desirable to process electrically before recording, either by special circuits or by limitation of the frequency response of the recording equipment.

There are two main factors which operate in deciding what type of recording equipment to use. Firstly is the frequency response required to obtain all significant data from the type of test envisaged; secondly the number of channels of information which have to be recorded simultaneously. Either of these may become an overriding priority in the selection of the recording method—or there may be no special preference.

Only a few years ago, if frequencies ranging to 15 kHz had to be recorded, there was no real alternative to high speed oscillographs recording on fast emulsion recording paper. At this time such devices performed yeoman service, but they have a signal disadvantage in that the recording paper has to be developed, fixed and dried in lengths up to 300 ft. It has also to be handled in darkness, or a photographic safelight, until fixed.

With the technical improvements, made in recent years, in ultra-violet lamps and specially matched recording papers; together with the vast improvement in galvanometer frequency response, the u.v. multichannel recording galvanometer has virtually superseded the multichannel oscillograph. The changes brought about by this revolution in technique have made enormous improvements in the number of channels available; in addition the recording requires only a few minutes' exposure to light after which it can be immediately examined and analysed. Figure 18 shows the improvement obtainable from a modern u.v. galvanometer compared with older oscillograph equipment.

Because the recording paper can be handled in normal lighting (with a few simple precautions) the width of the recording paper is not now limited, as it was with the oscillograph recording cameras; nor is the number of channels restricted by spacing of oscillograph tubes.

Improvements in galvanometer suspension and design have raised the top frequency response to about 15 kHz, though at a sacrifice of sensitivity which involves the use of driver amplifiers. Collateral improvements in multichannel tape recorders and tape have also helped. If now it is required to record frequencies above 15 kHz one simply makes a tape recording in real time at, say, $7\frac{1}{2}$ in, per sec and then replays into a u.v. galvanometer at half or quarter speed, making the necessary correction to actual frequency after analysis.

The results of this silent revolution described above, mean that frequencies up to, at least, 50 kHz can now be dealt with on a continuous basis. The rather inconvenient cathode ray oscillograph ' drum ' camera is now only required for very high speed events, where its very limited length of recording can be offset by suitable triggering from the event to be recorded—such as explosive phenomena and the like.

As indicated earlier it is now usual to tape record most of the testing time, even when recordings are not necessarily needed; in addition to direct recording on e.g. a 50-channel galvanometer, those parts of the test which are known, or expected, to be of direct relevance to the test. In general the only testing which would be undertaken, without direct recording available, would be ' flight ' tests of engines and tests made in the field on a casual basis. In these cases tape recording alone would be used.

There are a few less usual types of testing which require a special approach; notably in the use of blade tip passing probes (see Chapter

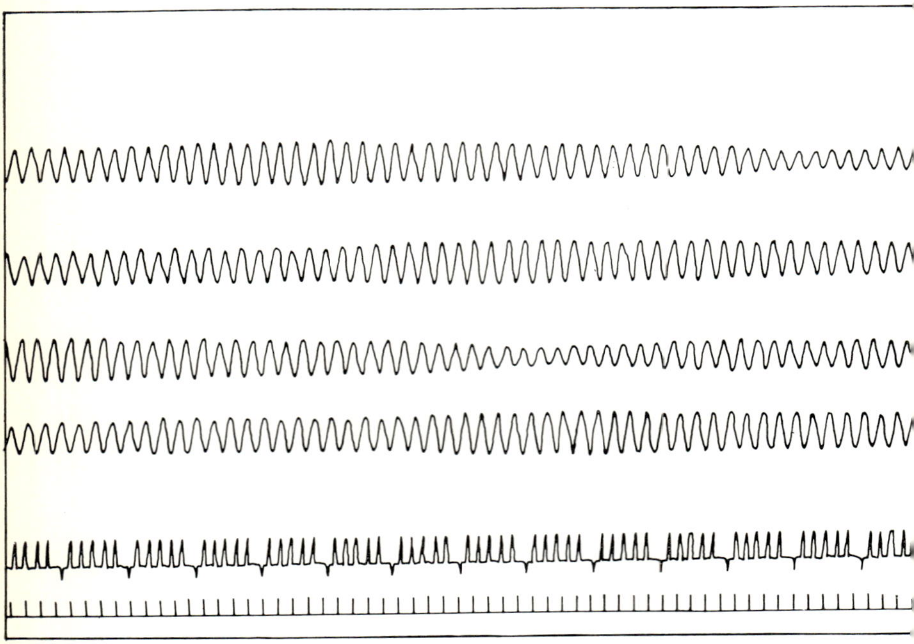

Fig. 18. A comparison of a recording made on an olde
6 channel oscillograph recorder with that possible o
a modern 50-channel U.V. galvanomete

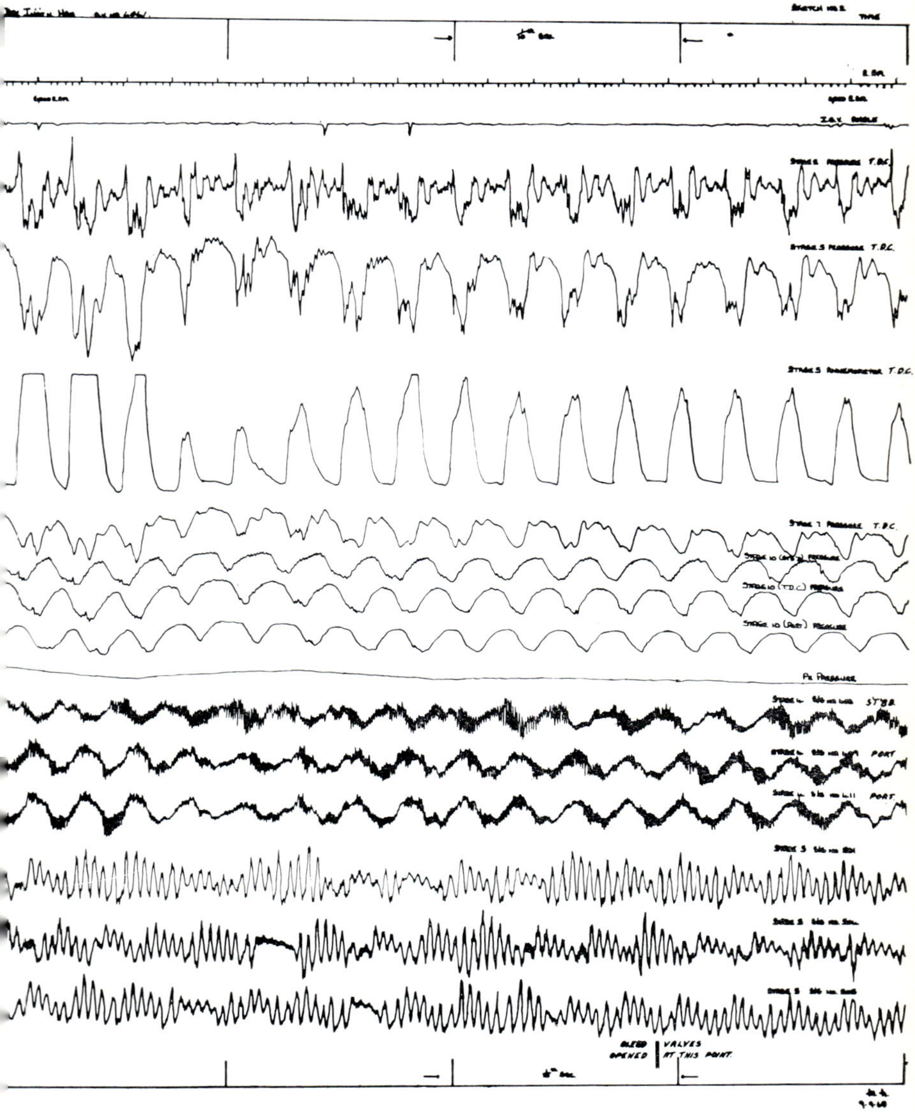
VALVES
AT THIS POINT.

7.10), in compressor or turbine rotor vibration measurement. At present these methods take the output from stationary coils, from a Doppler radio probe or a fibre optics ' lightpipe ' and require to be displayed as a two-dimensional spatial picture on a television screen. At the present time such information cannot easily be tape recorded and it is usually observed visually and recorded with a cine or Polaroid camera.

CHAPTER 9

data reduction, analysis and replay

As indicated in Chapter 8, a great deal of information is obtained from many tests while they are proceeding. However, it is unusual to find that sufficient data is so obtained for the solution of complex problems, while much testing time can be saved by undertaking the full reduction of data away from the actual test.

Some overlap between this and the previous chapter is inevitable, because all the methods described here can, if necessary, be applied to tests while the engine is running; although to do so may well multiply the cost several times. Also there is the problem of space for all the gear, necessary for various types of reduction, at the test location itself. Test beds are generally some distance from the other offices and facilities, because of the noise involved; and also it may be much more convenient to deal with the more involved replay operations at base—where supervision and personnel are more complete.

In the case of ' flight tests ', where instrumented engines are fitted to an aircraft, it is unusual to find that there is room for more than a bare minimum of recording gear in the aircraft; hence a compact battery operated multichannel tape recorder is all that is normally flown—apart from the necessary polarizing and preamplifier supplies. In this case data reduction remote from the test is mandatory. Plate 30 shows a seven channel compact battery operated tape recorder, such as would be used for flight data storage.

Until a few years ago the information from flight would have been telemetered from the aircraft to a ground station. However, the analysis of such results is more involved, and less complete, so that now effective tape recorders are available with the necessary number of channels and performance, the use of these is preferable.

Circumstances militating against telemetry are the effects of radio interference, while the restriction on the number of simultaneous channels implies the use of time multiplexing, with the consequent discontinuity in information in each channel. Because of changes in the signal path, together with the possibility of multipath reception, calibration must be continually transmitted—if amplitude modulation is used. On the other hand if frequency modulation is used to provide continuous data from all channels, on a v.h.f. or g.h.f. band carrier, the necessary band width may cause problems of a domestic, political or operational nature. Only when the return of the test vehicle is uncertain, as in rocketry, would one prefer nowadays to use telemetry.

1. *Visual observation of signal replay*

Although a warning was given in Chapter 8.3 against the possible dangers of electrical filtering of the signal before recording; it is clear that no such objection applies once the unfiltered recording is made on tape. During replay one can apply any combination of a number of signal processing procedures to the output from the tape replay, and repeat this as often as interest or time permits.

PLATE 30. A portable seven-track battery operated tape recorder such as would be used for outstation or flight recording. Solid-state electronics alone make such a compact device possible.

It will still generally be necessary to monitor on to a cathode ray oscilloscope for visual observation, so as to discover what sorts of information and interference are present. During this phase of the replay decisions can be taken as to which sections are to be recorded, and in what manner these are to be preprocessed.

It is but rarely possible to record all that is on the tape—and with most tests very wasteful also. A tape recording of a set of flight tests may well last one or more days (in real time) during most of which time nothing significant is happening. During the visual replay the operator will have the ‘ shot sheets ’ in front of him, showing what manœuvres are being undertaken, together with observations on weather and other ambient conditions.

In addition there will be speech records, on one or more tape tracks, giving test engineer's or pilot's comments on what is happening. Sometimes these are a bit lurid if things go awry and would need censorship before playing to a mixed audience! Spoken captions to each recording are a must and often provide the only sure link between testing and data reduction personnel. Such captions will show what calibration frequencies and amplitudes are used, which transducers are taken to which channels and similar information—indeed all detail relevant to the test should be spoken on to the tape; then if the test sheets take off in a high wind, or get rained on, all will not get lost.

Another important item, included here from bitter personal experience, is to have adequate reference, by shot sheets or on tape, between test engineers of the instrumentation team and those operating the engine. Once in a while such records prove vital, especially when an engine or rig failure has taken place during testing.

During visual replay there is an opportunity to settle a number of important points, and to obtain an approximate idea of the magnitude of any interesting peaks of stress or other parameter. This is also the time to check the synchronism of stress or other effect with engine revolutions, since it is vital to know if the effects found are excited by multiples of engine order. During the replay on to the monitors, decisions will also be made whether to replay in real or false time (i.e. faster or slower than the original recording), in order better to resolve very high or very low frequencies. The customer can with advantage participate at this stage.

In the case of a number of quite simple tests, no more than the above visual examination may be necessary; in most cases, however, decisions will now be taken to replay relevant sections on to recording paper in a u.v. galvanometer; or into a spectrum analyser or other type of recorder. It will also be clear what channels require filtering, integrating, differentiating, or perhaps pulse sharpening—in the case of such events as speed signals whose time of occurrence needs precise definition. Plate 31 illustrates the type of facilities needed for replay of complex engine tests.

2. *Graphical recording*

Some tests may be more easily assimilated if fed into X/Y plotters, two parameters at a time. Here because of the limited response rate of most crossplotters, it will often be necessary to replay at very low speeds. It must be remembered that calibrations at relevant frequencies are especially important when replaying from direct recordings in false time. The output correction circuits of tape recorders are naturally so arranged as to compensate, for recording and replay losses in frequency response, when both are made at the same speed. Such

Plate 31. A general view of a data reduction room, fully equipped for replay from tape into a variety of recording units.

restrictions do not apply when f.m. recording is used, though pulse information may need reshaping (sharpening) if its time definition is critical.

When replaying into u.v. galvanometers, or less often high speed pen recorders or cathode ray oscilloscopes, the same precautions are needed as for direct recording on an actual test. Suitable chart speeds will be chosen; on the other hand if the choice proves to be unsuitable it is easy to go over it again in the replay room. It may be desirable to record at more than one speed; for instance one recording may be made at quite a low speed, insufficient to resolve the frequency of the signal but showing its amplitude throughout a rather long time; with other short higher speed recording at intervals, only to confirm the frequencies present.

3. *Analysis by graphical methods*

Leaving aside the special case of crossplots, which are self-evident, most of this work is conducted on continuous paper recordings from u.v. galvanometers today. The first step is to edit the recording, with a quick run through by an experienced engineer, when those portions to be analysed in detail are marked and the instructions are laid down as to the measurements to be made. After this the work of determining the frequencies and amplitudes can be done by less experienced analysers, who will at the same time separate the datum changes from the alternating ones, where both have been recorded.

When these figures have been corrected for the sensitivity of the gauges, or other transducers, and the gain of the amplifying channels, according to the frequency of the signal at the time, they are set out either in tabular form or as graphs of various kinds. It is at this stage that a considerable amount of judgement is required; given a large number of results from straingauges, and/or pressure and speed transducers, together with time references, it is vital to present these in a manner which is easy for the customer to understand at a glance.

A complete tabulation of the results from only perhaps 12 straingauges, 4 pressure transducers and 2 engine speed signals—which would be one of the rather simple sorts of test—would undoubtedly embody all the results; it would almost certainly be quite useless in helping to understand them! It is natural to plot graphs for interrelated quantities, the test was undertaken for just this purpose, but it may well be necessary to try several methods of plotting in order to make the relationships as clear as possible. In aero engine work a majority of results of straingauges are plotted against engine speed, thus to show the most harmful conditions quickly; however, if the stresses are caused by surge effects the relation of other quantities than speed will be important, and the relation of total air flow, or swirl vane angle or some similar variable would be used for one of the graph axes. Only occasionally are three-parameter plots required and, if they are, may be more easily made using a Visicorder—as dealt with in § 4 of this chapter.

Perhaps the main reason (other than that of the cost of the extra gear required for automatic analysis) for continuing with graphical analysis lies in the manner in which our own brains function. Presentations on the basis of Cartesian coordinates seem admirably to suit the marked character recognition bias which is inbuilt in our present thinking; true it may be more due to our school training, but it is there all the same. For many people—if not the large majority—time is like a never ending stream and flows from left to right! No doubt it goes differently in China, but to us it seems natural for events occurring in changing time to have that as the x-axis and the effect itself on the y-axis.

One of the great advantages of recording the results—at least at first—graphically is that switch clicks and other forms of electrical interference are easily recognized and can be removed from the record when this is realized. Another benefit of such recording is that time modulation of amplitude (periodic or cyclic variation of the height of the trace) is recognized more easily than with many of the automatic systems. Also the graphical recording will show whether signals at two different frequencies alternate or coexist. Figure 19 shows a quite simple example of time amplitude modulation.

Airblown self-excited vibrations of blading, such as are excited by unsuitable angles of attack, or surge, often show the former effect. The signal may alternate from, say, second flap to first torsion or the like. In rotating stall excitation of compressor blades, although the stall cells do resonate with the blades the number of cells changes randomly giving high amplitude modulation of the straingauge signals. There is a marked difference in the behaviour of the blades, in fatigue, depending on whether the stress is modulated or unmodulated. All these effects are obvious on recorded graphical traces, while they can too easily be missed by other methods—especially if these methods are essentially time averaging in nature.

Although the advantages of full graphical recording have been rather heavily emphasized above, they are without doubt time consuming, tedious and generally contain much more data than is ever recovered, though some of this may not be of significance. Where the more sophisticated automatic methods are used it is usual to examine, at least the more critical portions of the data, by graphical recording as well.

4. *Automatic analysis*

One of the simplest cases of the use of automatic analysis is when one channel of engine vibration has to be monitored. Here it is easily possible to decide beforehand whether to concern oneself with displacement, velocity or acceleration. This done the results can be rectified and displayed as a meter reading. As long as it is reasonably certain that there is only one significant cause—e.g. engine unbalance, then no further clarification required. If, however, it is necessary to know which other frequencies are involved the meter no longer gives all the necessary information. So if we are faced with abnormal engine vibration we shall have to make ' on line ' spectrum recordings; or tape record or make graphical recordings.

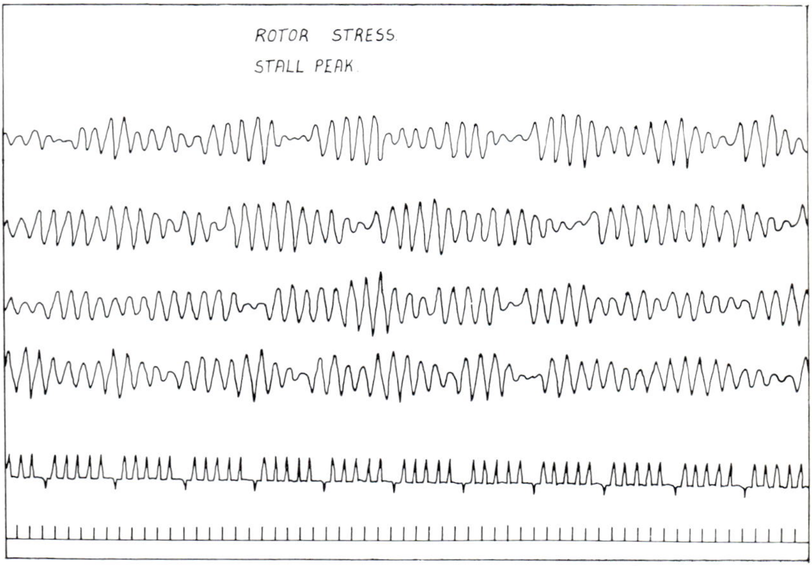

Fig. 19. A recording of four channels of straingauge signals, with engine speed and time signals at the bottom. Note the manner in which the stress fluctuates, a typical effect when an engine compressor is in a state of rotating stall.

Similarly, with the more sophisticated analysis systems to be shortly described, if any doubt arises as to the integrity of analysis made by automatic methods, it will be prudent to check the tape output, visually or by short galvanometer recordings.

However, high speed spectrum analysis by such as the Federal Scientific Ubiquitous Spectrum Analyser can save an enormous amount of time, or alternatively provide a great increase in data obtained in the same time. For the best results it is desirable to combine this equipment with the Honeywell Visicorder. By the use of these systems it is possible to extract an immense amount of information, compared with graphical methods, in an absurdly short space of time—if need be while the engine is running!

The Ubiquitous Analyser is a sophisticated analogue to digital to analogue spectrum sampler, which uses computing techniques to produce an extremely rapid yet accurate spectrum of the incoming signal and permits working ‘ on line ’ to the engine if required. However, if the engine signal itself is changing, or cycling, then one spectrum at a single time is not complete information and the analysis must be repeated at short intervals—either of engine speed or time; thus building up a picture of the change in composition of the signal. Such information can be presented to a recorder capable of a two dimensional display—such as a cathode ray oscilloscope—with the successive tracks stacked one above the other, so as to form a sort of contour map.

An example of this type of presentation is given in fig. 20, which was produced on the Honeywell Visicorder. This is a refined and comprehensive oscillograph, with a fibreoptic screen, which enables exceptionally good definition to be obtained on sensitive film or paper, without the need for a camera. It is also equipped with the facility for *Z* (beam brightness) modulation; which for certain purposes is more convenient than the stacked shot illustrated, but lends itself less to reproduction in a book.

Figure 20 is a stacked shot trace which reveals a wealth of information to the experienced eye. It covers the individual spectra, produced at intervals of a fraction of a second, while an engine decelerates from top speed to idling. The stress amplitudes shown are from a rotor blade straingauge and the apparent height of the ridge is proportional to stress. The ridges, from left to right, represent excitation at 1 EO; 2 EO; 3EO and so on. Where blade modes coincide with these the ridges appear to rise into peaks. Rectangle B–B shows (a) how the 350 Hz mode (at idling) pulls into and runs with 3 EO and (b) how the 1500 Hz mode resonates in turn with 10 EO, 11 EO and 12 EO. The lower shot shows a time averaged spectrum at A–A in the top speed condition.

While such elaborate equipment is very successful indeed in breaking down each of the effects combining to give the total output of any transducer, they do not always give the full value to the total stress, or other effect. When heavy amplitude modulation of the signal is present the automatic devices do not necessarily reveal the full composition with time or the absolute peak condition. They would then need to be supplemented by a peak reading voltmeter, preferably associated with a reversals counter operating at a number of different levels.

An alternative to the last comment is to supplement the automatic readout with a slow speed galvanometer trace to give the amplitude of total signal. In this case the paper speed could be so slow as to preclude any recognition of separate cycles, this information coming from the spectrum analysis. If the results are to be used to programme a random stress fatigue test then full statistical information could be obtained from the automatic equipment. The writer would prefer, in such a case, the much simpler way of recording on tape a ‘ potted ’

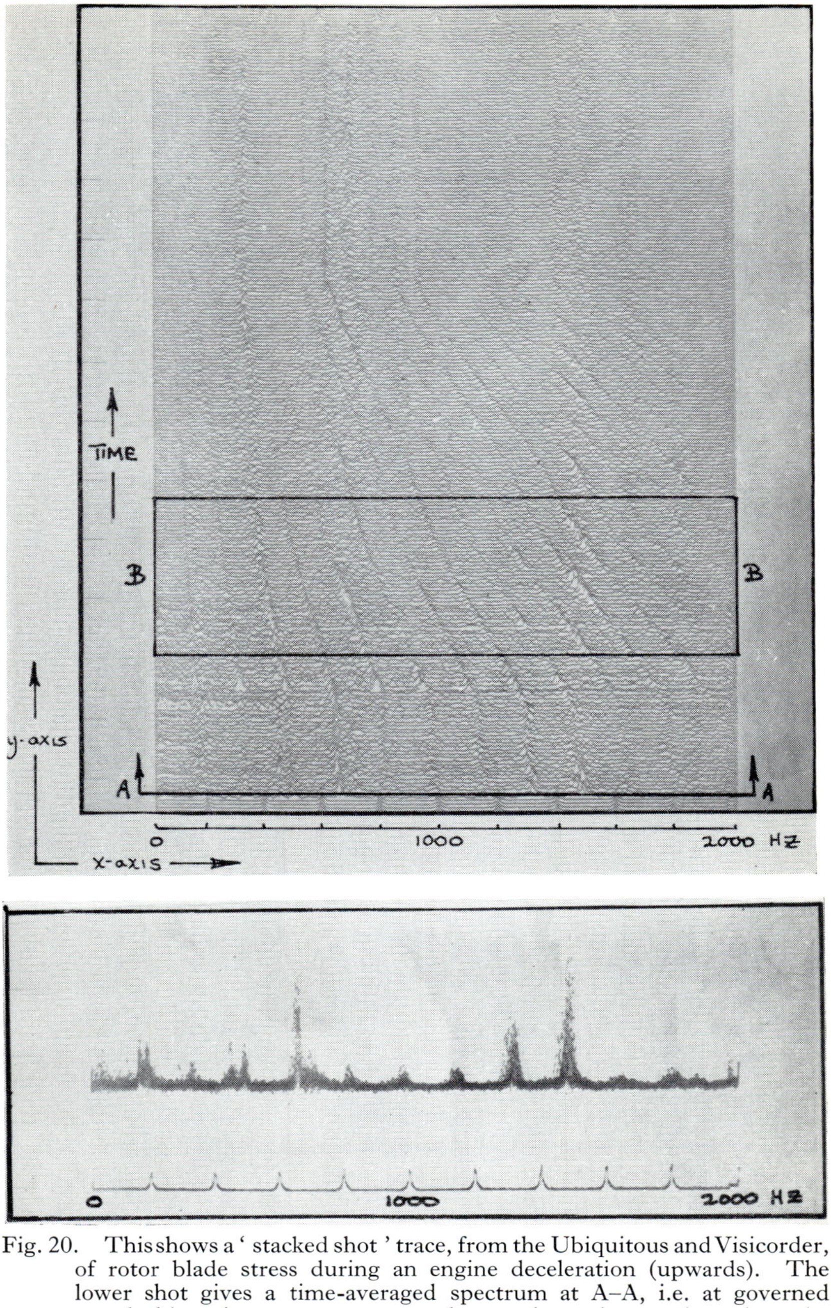

Fig. 20. This shows a ' stacked shot ' trace, from the Ubiquitous and Visicorder, of rotor blade stress during an engine deceleration (upwards). The lower shot gives a time-averaged spectrum at A–A, i.e. at governed speed, this assigns more accurate values to the peak stress in each mode.

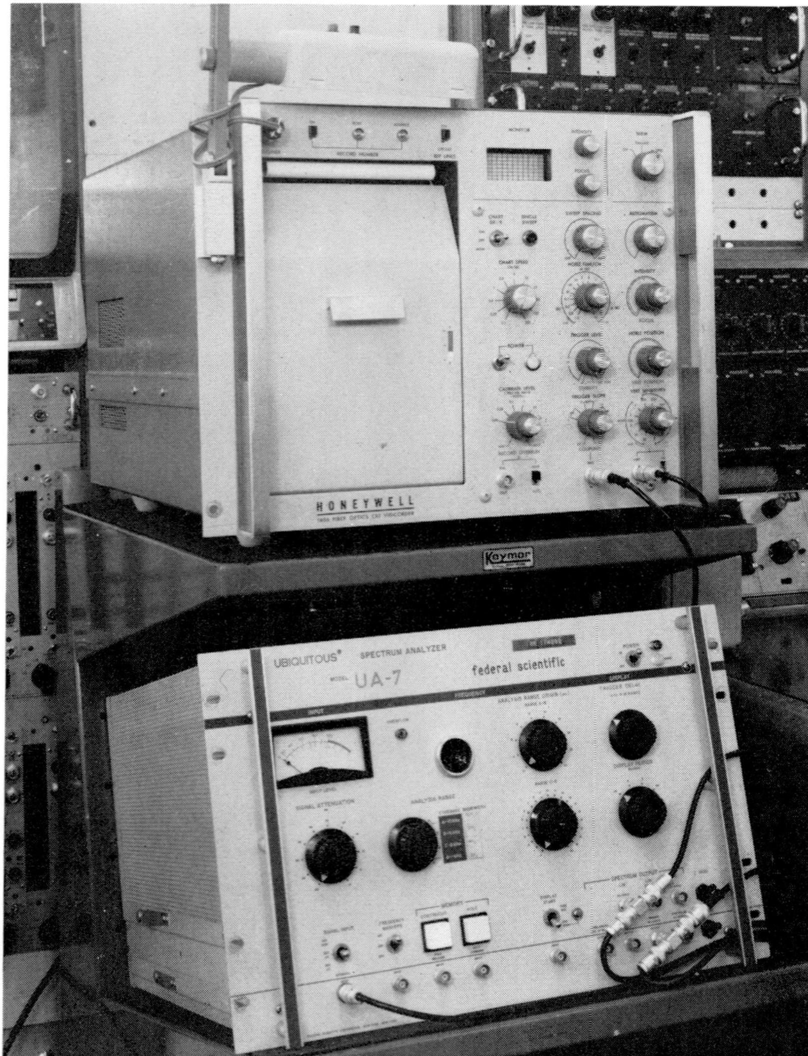

PLATE 32. This illustrates the sub-unit, of a replay facility, comprising a Federal Scientific Analyser and Honeywell Visicorder cathode ray tube with fibreoptics screen.

flight record from appropriate straingauges and using this to programme the test.

It is pertinent to observe that the emphasis on a knowledge of the full composition of the stress pattern of a blade, or other component, is of greatest importance when dealing with engines in the initial stages of their development; in the case of engines in full fleet service one is more often looking for a particular circumstance of operation, which gives a single mode of excitation, than of searching for many different ones. In general, if there were many different significant excitations it is unlikely that the engine would ever have reached fleet service.

The automatic system is at its most useful when the engine is at so early a state that it may not even be safe to run it steadily up to full speed, without a good deal of data on the blades; because its operational limits, and the best settings of the control parameters will not yet have been established. In these cases the very high cost of the equipment is fully justified. Vibrations of casings, fluctuations of pressures and the like may well be better dealt with by the simpler graphical methods. Indeed the simpler the modes of excitation the less need is there for complication of the investigating gear. Plate 32 shows the Ubiquitous Analyser and the Visicorder.

CHAPTER 10

sound room techniques

FREQUENT reference has been made throughout the text to 'sound room tests'—chiefly in contradistinction to engine and other rig tests. It is appropriate in this chapter to discuss such methods in more detail. The phrase 'sound room test' is not, *per se*, meaningful, outside the ranks of acoustic and vibration engineers, yet within these circles is quite clear in scope.

By such a term we mean (literally) tests made in a sound room or 'deadened' room. It is not generally necessary to have a room so absorbent as to produce no reflections from its boundary surfaces, i.e. truly anechoic; but to reduce the conditions inside to such a degree of quietness and freedom from vibration that external sounds do not intrude, nor do the sounds generated within disturb others in the neighbourhood. Plate 33 shows a corner of such a room.

The reasons for making natural frequency tests in such an environment are twofold; firstly in order to define these frequencies accurately it is important to obtain conditions of low damping which means sharp resonances, these can be piercing to the ear in the range above, say, 2 kHz; secondly because the means of detection would be adversely affected by other noise or vibration. In many cases the ear is a very good detector of resonance by itself, though other aids have often to be enlisted to define which particular mode is being excited. Microphones, accelerometers and proximity probes are common detectors; while violin bows, electromagnets, piezoelectric crystals, moving coil exciters with coupling probes, and small loudspeaker units are all used, from time to time, as means of excitation.

The philosophy behind this type of work is to relate the frequencies at which parts resonate in an engine to the type, or mode, of deflection experienced by the components. It is rarely possible to use enough transducers for an engine test as would be necessary to define completely the mode of vibration of a given part, but the frequency of vibration is easy to obtain from the observations or recordings. A set up for checking the vibration modes of a centrifugal impeller is shown on plate 34.

Given the frequencies of any resonance of a component, it is a much simpler matter to investigate the modal shape apart from the engine. While doing this it is also not difficult to check what modifications might 'tune' the resonances to different frequencies. Ideally the

Plate 33. A corner of a small investigation sound room. The particular job being checked is the sun wheel from a gearbox, the end masses simulate the relevant panels of the gearbox. The mode under investigation is the bending of the gear in its spring cage bearings.

sequence which should be followed in investigating failures in an engine from resonance caused by fatigue is as follows. First, build a test engine instrumented so as to measure the harmful stress or vibration, from these determine the frequency and magnitude of the resonance, then do some sound room tests to define what modes are being

PLATE 34. Checking diametral modes of a centrifugal impeller. As will be evident quite a small moving coil unit is ample to give sand patterns. In fact, since the impeller was enamelled dark brown common table salt was used for sand, to give a better contrast!

excited. From such results it should be possible to take suitable action to avoid further trouble; either by reducing the excitation or removing the resonance out of harm's way, in terms of engine speed or other conditions, or by modifying the part to change its frequencies.

On new designs the above course of action is not possible, and in this case sound room tests may be commissioned to determine the resonant frequencies of a new design of component and find if any of these are likely to be excited in the running engine. In many cases this looks like crystal ball gazing; however, there are a number of known excitations within the engine which are likely to be important and resonance with these, at any condition which the engine is likely to maintain for much of its life, should be avoided.

The various modes possible have been dealt with in earlier chapters of the book, so there is no need to detail them again. Instead, rather, this chapter deals with how to obtain clear and accurate results from, in the main, simple components and assemblies.

One difference between the running engine, and such testing, as is here described, is the effect of damping, associated as it is with power and force required. As already noted Q values in the running engine may well range from 5 to 50, and only seldom much higher. On the other hand large powers are circulating within the engine because the turbine drives its own compressor, besides producing output power. A medium-sized jet engine can transfer as much as 100 h.p. (75 kW) from each blade of an early compressor stage; if only 0·10% of this is converted to vibration energy this would be 75 W, which with a Q of, say, 20 would be easily capable of failing the blade in resonant fatigue!

With careful attention to clamping and mounting much smaller powers can be effective for resonance searches, with a consequent sharpening of the response and greater accuracy of frequency of the peak. A 15 W domestic hi-fi amplifier is quite sufficient for resonant searches of complete turbine rotor assemblies, and often whole engine carcasses. For single components much less will suffice; the power a violin bow can provide is only a fraction of a watt, yet quite adequate to excite sufficient movement in a blade to provide clear sand patterns of the nodal systems. The ease of frequency definition and the related forces needed are illustrated in fig. 21 in relation to the Q achieved in clamping.

The writer would remark here that when making fatigue tests of components, by resonance in natural modes, higher powers are required —nevertheless attention to clamping, to secure a high Q, reduces the power and force capacity required by quite large factors.

It is worth noting that a knowledge of music—and in particular of how musical instruments produce their sounds—is a great asset in sound room work. The ability to play instruments, particularly one of the violin family, is also helpful. One of the most successful technicians known to the author in this field (a violinist of some experience) could produce the diametral modes of a 16 in. diameter centrifugal impeller up to the fifteenth overtone with a cello bow alone!

A trained musical ear, such as is possessed nowadays by many hi-fi enthusiasts, as well as by practising musicians, is also of great value in quickly determining resonances, and in detecting such cases where overtones and fretting are present. Successful sound room work is still as much an art as a science, and there are many tricks found by experience, which the author hopes will save readers from finding out the hard way.

1. *Types of vibration measured*

The natural and resonant modes of vibration possible for engine parts were dealt with in the first half of the book. The most common requirement, in gas turbine work is perhaps that of dealing with flap, torsional and edgewise modes of turbine and compressor blades. Discs

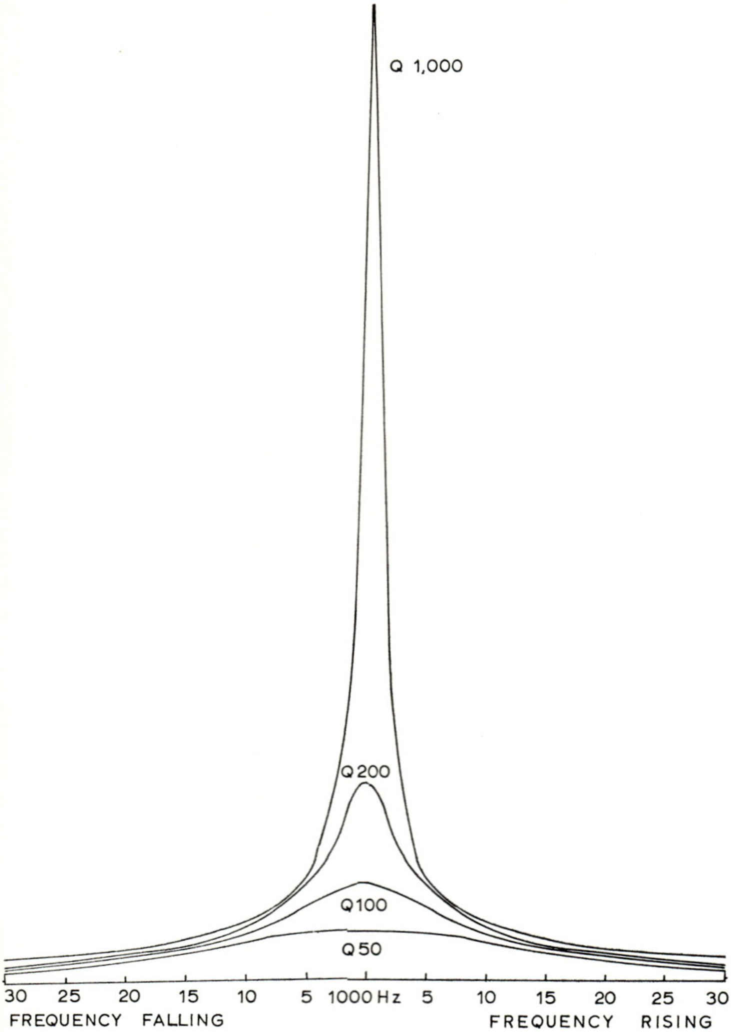

Fig. 21. Plot of response to a constant force over 30 Hz either side of a 1000 Hz resonance for *Q* varying from 50 to 1000. *Y* axis shows deflection in arbitrary units. *X* axis frequency deviation in Hz.

of compressor or turbine are frequently tested, and in many cases it is necessary to make such tests with the blades attached. In such items the modes of the blades are modified by their coupling to, and through, the disc ; and disc modes by the presence of the blades, as they would be in the running engine.

Pipe and wire vibrations require the relevant casing components, so that these components may contribute their attachment stiffnesses correctly. For these components bowstring modes in several planes may be required. If harmful resonances have to be altered, various clipping points may be more easily investigated in the sound room than on the test bed and much more cheaply too!

Gear diametral measurements are sometimes required and if these must be altered it is helpful if the various alternative designs are all available for test at the same time. It will not generally be necessary for the teeth to be cut. Sheet metal parts, such as combustion cans, jet pipes and such like will be tested for diametral (or bell) modes and sometimes air cavity resonances.

Accessories will require testing on their mounting brackets; in this case as with pipes, it is often necessary to make such tests on the casings to which they are attached; since the latter frequently contribute as much of the flexibility (compliance) as do the brackets themselves—at least with the light casings needed for aero engines.

When investigating torsional resonances of rotor assemblies it is often possible to use flywheels which correctly reproduce the polar inertia of, say, a turbine assembly, though often it is quicker to use the assembly itself than wait for flywheels to be made.

Hydraulic vibrations will almost always call for a total assembly of the relevant pipework, control units, recuperators, etc. for a satisfactory determination. The remarks made earlier in the book about priming, i.e. getting rid of all air, are particularly important for this class of work.

Since a high Q is needed for accuracy—as well as to reduce power requirements—damping in assemblies is very undesirable. Where, in the running engine, splined drives are held in firm contact by torque this must generally be simulated in the sound room either by making such parts an interference fit—or more simply and equally effectively by using Araldite or similar adhesive to glue the parts together. If this is not done and tests are made with a ' rattling good fit ' not only will the frequency band of the resonances be broadened by an intolerable amount, but also the frequency of such resonances will become dependent on the amplitude—and generally also erratic.

2. *Methods of mounting*

If any reader should doubt the need for near perfect fitting in sound room testing, let him clamp a compressor or turbine blade in an ordinary fitter's vice and pluck it. It will be found to be as ' dead ' as a piece of lead or plasticine. Properly mounted such a blade will go on vibrating for several minutes, in the vice it will execute a few cycles only. The main damping here will be found to be fretting of the vice handle in its hole!

In general there are two broad alternatives in sound room mounting; either give the part complete freedom or '100% clamping'. Complete freedom can be secured by hanging a part from Bungee rubber or supporting it on foamed plastic. A major disadvantage of this method is that it is difficult to hold most components stably in this way and that the part will probably move relative to the detector, and sometimes the exciter used. It is, however, quite suitable for disc investigations using a bow to excite and the ear to detect. When measuring such components as compressor or turbine blades, which are 'encastered' to a greater or lesser degree in the engine, it is clearly inadmissible to use this method, since it does not simulate engine conditions.

If complete freedom is inconvenient, or incorrect, then '100% clamping' must be aimed at. Where the component has a suitably machined root fixing or lug then it may be clamped in a vice or fixture—provided this latter has no loose parts and that the 'fit' at the clamped junction is really good. This means a quality of joint which would satisfy a skilled millwright. Often it is necessary to use 'marking blue' to check for a good fit.

Disc type components, if mounted by the hub in the engine, may be clamped between substantial machined washers on a rigid pillar and base. Again the base, often conveniently a surface table, must either be rigidly attached to the floor (its weight generally secures this) or be mounted on resilient mountings, the latter is generally to be preferred as it also prevents floor borne vibrations confusing the readings. Spanners and other tools lying loosely on such a surface table are often responsible for a mysteriously low Q.

When the tests involve 'tuning' the component to a different frequency by removing or adding mass between tests it is particularly necessary to secure consistency of clamping in both cases. A useful guide here is that anything which reduces the stiffness will drop the frequency of any given peak. If alterations to clamping raise the frequency of any mode, then that which gives the highest value is the most nearly correct.

It may well be argued, especially in the case of rotor blades, that the frequency of modes as determined in the sound room is not that which will occur in the engine and this is strictly true. However, for stationary parts there is little difference; while the values determined statically for rotating parts, can be corrected by computation with much better accuracy than if the frequency of the various modes in the engine were computed *ab initio*. Also the shifts in frequency determined statically will not be very different *pro rata* for the engine case. Plate 35 shows a compressor blade being checked for resonances.

As an idea of what can be obtained in the way of adequate clamping, a rotor blade with good root fixing in a parallel, or angled, vice should have a Q between 200 and 1000, while a turbine disc may have as high a Q as 8000 in diametral modes.

A problem arises when it is required to determine the frequencies of blades (particularly welded pairs or triples) with firtree or other complicated root fittings—which are not easy to hold effectively. A

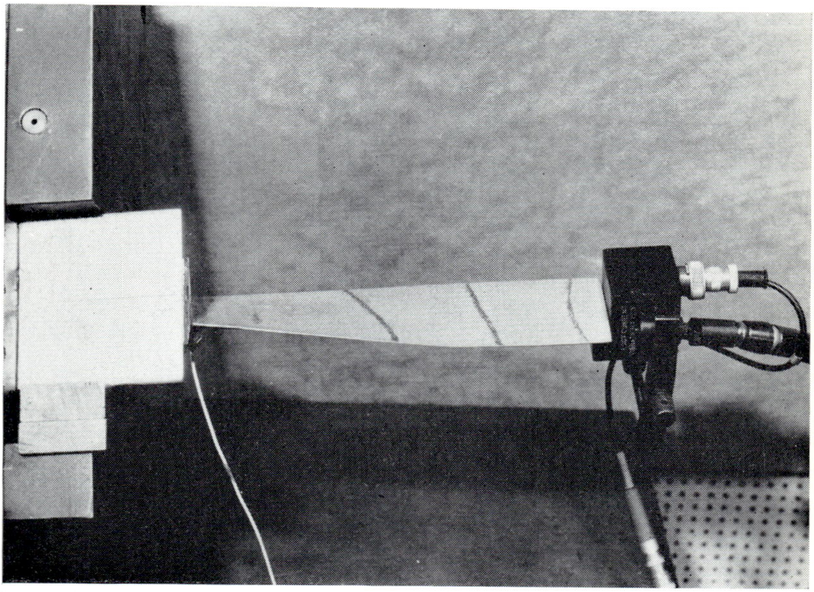

PLATE 35. An aluminium alloy compressor blade being excited in fourth flap with an eddy current magnet of Rolls-Royce manufacture. Note the small peizoelectric accelerometer on the root of the blade to assist in tuning to the peak.

method which the author has used for many years can be very useful in such cases, and as it may help others faced with the same problem is described here. What is required is a metal to cast round the component, to the depth of the encastre required, which will grip tightly, be acoustically live, strong and easily machined, but not melt at so high a temperature as to alter the component in respect of its structure.

An alloy which meets the above requirements is a zinc-based formulation, with small amounts of copper and aluminium. Two proprietary examples are ‘ Kirksite ’ made by Hoyts Metal Ltd. and ‘ Kayem ’ manufactured by Imperial Smelting Co. Ltd. The metal was originally developed for making short-run dies for sheet metal pressings and can be found in the tinsmiths’ shop in many firms, where it is used for mandrels and formers, as well as temporary dies.

The high expansion coefficient (linear) of 28×10^{-6} degC means that it shrinks, from casting temperature downwards, on to steel, titanium, aluminium and indeed almost all other metals. It melts below 400°C, is acoustically ' live ' and machined easily to a good finish, almost as well as freecutting brass. In addition it has sufficient compressive strength to allow really high clamping pressures. In fact, the hold which this alloy maintains is so good that resonant fatigue tests on nickel chromium alloy turbine blades can be carried out with the firtree roots cast in zinc alloy without loss of grip, or significant ' bell mouthing '. Plate 36 shows a firtree-rooted turbine blade mounted for frequency check.

The low melting alloys of lead, tin and bismuth—such as ' Cerrobend '—are rather ' dead ' acoustically and will not stand sufficient clamping loads for effective encastre of most engine components.

Often too little attention is paid to the mass necessary to ensure effective and repeatable terminating conditions. There is no hard-and-fast rule, which will fit all cases, but good general practice will be met by making sure that the static mass is at least a hundred times that of the component under test. This means that a standard heavy 2 ft × 3 ft cast-iron inspection table will do for most engine components. Such a slab can be easily carried on four stock vibration mountings, has the necessary flat surface and good rigidity.

The box angle or cube, so convenient with its numerous slots in all faces, should be avoided like the plague. True there is always a place where wanted to attach your component, but all the short beams between the slots have their own transverse resonances and different parts will resonate at all sorts of frequencies from 100 Hz upwards. For endurance rigs such conditions can be allowed for, but in exploratory work never!

3. *Excitation devices*

A prime consideration, in selecting the type of excitation for a particular job, is to use one which is reasonably effective without significantly modifying the resonances of the part under test. For example if a medium-sized electromagnetic exciter has its probe rod fixed to a normal-sized compressor blade the veriest tyro can see that the attachment of, say, $\frac{1}{2}$ lb of moving coil and rod to the blade will alter profoundly the very natural frequencies we have to measure.

Such an example is, of course, extreme, but it is for this very reason that so much stress has been laid on obtaining a high Q by the very best clamping. With a high Q only very small forces need to be used, and so there is little alteration to the part's own resonances by the driving device.

3.1. *The bow*

The author must confess to a predilection for the violin, cello or bass bow as the exciter least likely to modify the true natural frequencies

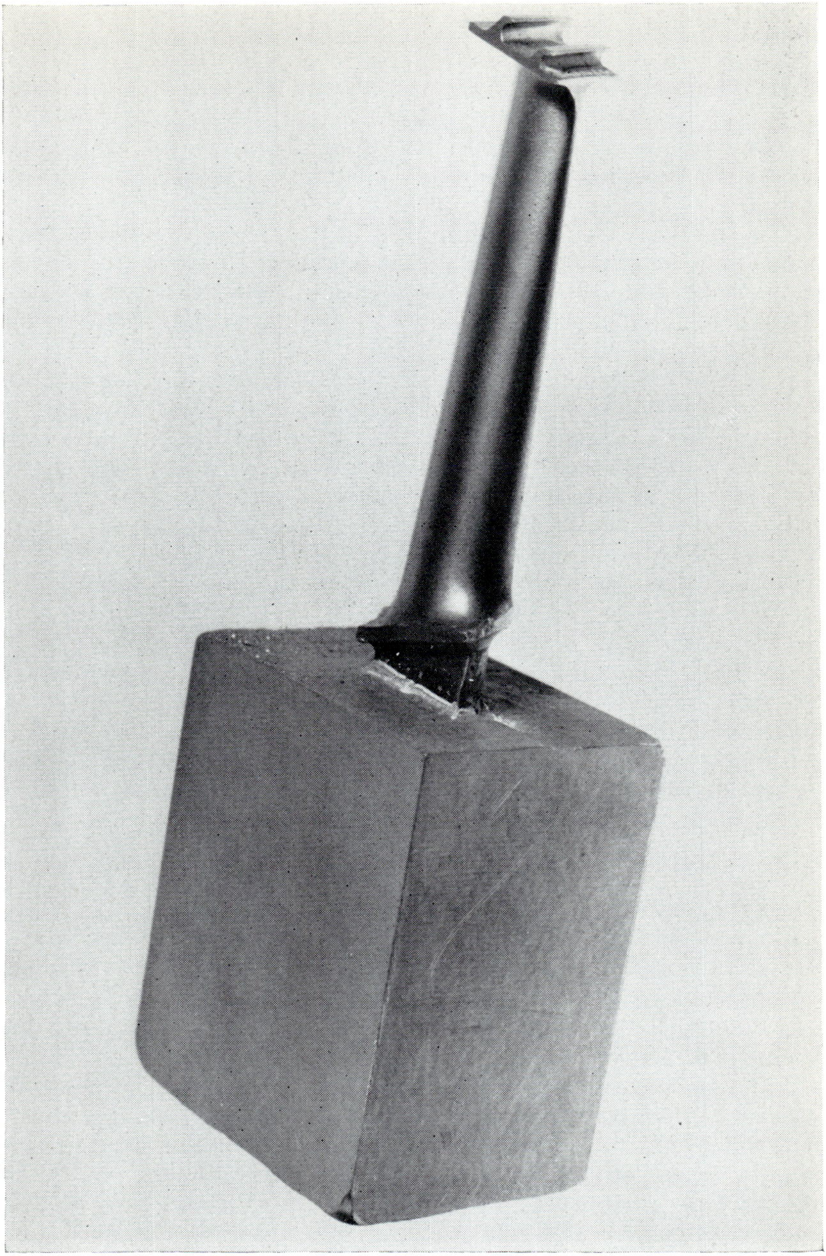

PLATE 36. A turbine blade with the firtree root cast in Kayem metal and two faces machined.

of a component; and would use it as a cross check whenever doubt arose.

The reason for this is that the bow motion is truly aperiodic and hence cannot easily couple to the part, though it is occasionally possible to produce artificial modes by certain bowing and stopping positions. If, however, the vibration continues, *at the same frequency and with the same nodal positions*, after the bow and stopping have been removed, the mode is a true natural one.

The stopping referred to is by placing the fingers at suitable nodal points so as to assist the production of a particular overtone, analogous to the light stopping of a harmonic on the violin, or other stringed instrument, at an appropriate aliquot part of the length. The method is illustrated in plate 37.

It is not unduly difficult to obtain the first four or five flap and torsional modes of a compressor blade and the first two or three from a turbine blade. Skilled string players can do much more, once they have made the transition from compliant string to rather rigid engine part, but not all engineers find bowing easy and the shape of many components do not lend themselves to this method; for such cases several other means exist as set out hereafter.

Note: All of the subsequent methods of excitation involve driving the part from a predetermined frequency, generally derived from an electrical oscillator, which can be varied over the desired range of frequency. Hence it follows, by definition, that the frequencies are resonant not natural ones, though with a high Q there is no practical difference in the frequencies determined. *Never assume that the part is vibrating at the frequency of the drive.* It may be vibrating at some higher, or even lower multiple, of the oscillator frequency or some combination of these.

3.2. *Electromagnet exciters*

A simple iron-cored, or even air-cored, electromagnet placed near a magnetic component will excite this into a small forced vibration which will build up manyfold when a resonant frequency is reached. There is no additional mass attached to the part to alter its frequency and this is a popular method of exciting steel parts.

With non-magnetic parts the method can still be used if the electrical resistivity of the component is low since eddy currents induced in the part will couple with the magnet and give appreciable drive. Aluminium alloys, bronze and beryllium copper are all low enough in resistance for this method to be used. Plate 38 shows Rolls-Royce-made eddy-current magnetic exciters with permanent magnet bias incorporated.

If a simple electromagnet is used there will be attraction at both the positive and negative peaks of drive flux, unless the part is magnetically

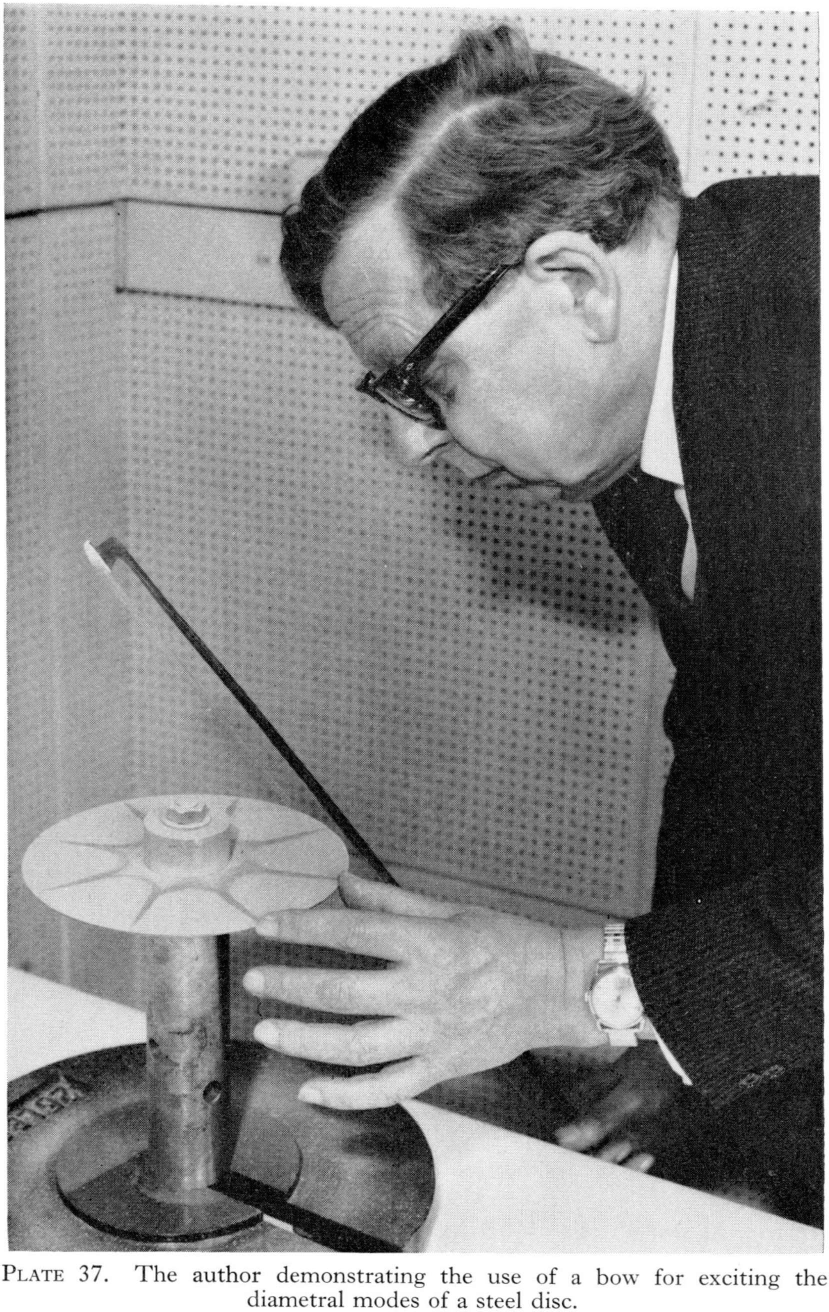

PLATE 37. The author demonstrating the use of a bow for exciting the diametral modes of a steel disc.

hard, and the component may run at twice the frequency of the oscillator. This can be avoided by inducing a permanent field, greater than the alternating flux, either by passing a d.c. current through the coil in addition to the a.c. drive; or by using a permanent magnet in the system, generally the latter. Magnetic exciters are available or can be made locally.

Because of high resistivity of titanium, nickel–chromium high temperature alloys and certain austenitic types of stainless steel, the method is not very satisfactory for these materials.

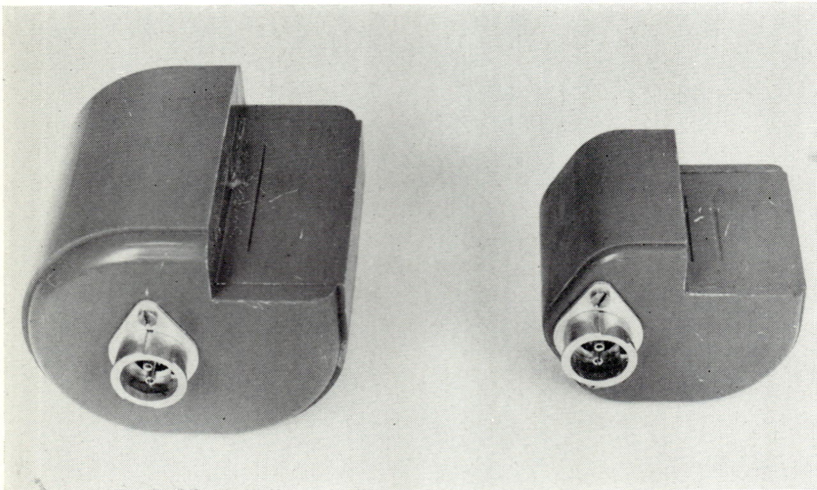

PLATE 38. Two eddycurrent magnets of Rolls-Royce manufacture. The laminations of the electromagnet are in one of the re-entrant faces and the pole of the permanent magnet in the other.

3.3. *Piezoelectric drive*

If a piece of quartz, suitably cut with respect to its crystallographic axes, or a bi-morph piezoelectric semiconductor is glued on the surface of the part it will produce a force tending to camber the part locally, and hence will drive the part. This method is most popular when a large number of determinations are to be made on one component, as in research work, demonstrations or when driving a rotating part in a spinning rig.

A major disadvantage of the method is the high impedance nature of the combination, which results in high voltages which tend to spark over the edge if the drive is too high, and requires matching transformers from most amplifiers ; which generally have output matching to low impedances of the order of 3 to 30 Ω.

Bimorph piezoelectric bender crystals of lead zirconate can be obtained no more than e.g. 0·5 in. long × 0·100 in. wide × 0·030 in.

thick so the loading on the part is minimal. If the crystal is cemented along one edge near the root of a compressor or turbine blade, almost all the resonant modes—even up to very high overtones—may be readily excited.

3.4 *Moving coil exciters*

This type of exciter consists of one or two coils moving in an annular gap with a strong magnetic flux in it. The arrangement is very similar to that of an ordinary moving coil loudspeaker, without the cone. In the smaller sizes the flux in the gap is always provided by a permanent magnet system, in some of the larger ones an electromagnet is sometimes used; however, such sizes are not normally needed for resonance searches. Plate 39 shows two popular sizes of moving coil exciters suitable for sound room work.

Plate 39. Two commercial moving coil exciters. The small one is of 5 W rating, the larger one of 350 W.

The moving coils are coupled to a rod with transmits the force from the coil to the job. The smaller unit illustrated is ample in output for blades, and many small discs and similar components, and the mass of the moving system is small; also the frequency response extends well above 10 kc. The larger one is suitable for turbine assemblies, jet pipes and the like and two of these may be used to generate an alternating torque for torsional investigations; the frequency response of the larger units is, naturally, inferior at the top end of the spectrum.

Still larger units are made, both electromagnetic and electrohydraulic, and these are used for such objects as airframes, ships and buildings. As such they are outside the scope of the present book. Occasionally, however, where it is necessary to test heavy accessories—perhaps 10–20 lb in weight—the larger vibrator, which will support the job without other assistance, may be convenient; though its frequency response is severely curtailed at the top end.

It is worth while, at this stage, to correct a misapprehension which is more common than it should be. This is that the effort exerted by the vibrator is related to the amount of movement of the coil assembly—this is not the case at all, a coil in a constant magnetic field will exert a force proportional to the current only.

This is particularly important when testing under such conditions that the job is at an anti-resonance with respect to the exciter, such as when small parts are mounted in a large mass and driven so as to produce resonance of the part against the mass. Under these conditions the mass scarcely moves at all but the force remains the same as if it was moving the full travel of the exciter.

It is important, when using moving coil exciters, to attach them as near a node as possible in order to minimize the effect of the moving coil mass on the frequency of resonance of the part. Thus for a turbine or compressor blade the vibrator would be attached quite near the root and certainly not at the tip. On the other hand attachment at a node itself will inhibit that mode, a dodge which is sometimes useful when measuring different mode families of a system which are close together in frequency.

3.5. *Air pressure exciters*

The title of this section is somewhat vague, there being a variety of terms used for this type of excitation. Basically what is meant is a device causing sound waves (pressure pulsations) to impinge on the parts to be excited.

While air sirens are often used for endurance and fatigue testing of sheet metal parts and blades they are of little use in exploring natural frequencies. They are clumsy to tune and do not cover a wide frequency spectrum.

The type of unit convenient for frequency checking is generally made commercially as a horn loudspeaker drive unit, such as is widely used in public address and cinema systems. The smaller sizes will accept from 5 to 15 W of drive power and larger ones are available up to 100 W power handling capacity. They do not, of course, deliver this amount of acoustic power into the air, being at the best about 30% efficient even when used with a suitable horn. Plate 40 shows two representative sizes.

As used for sound room exploration it is common to use a convergent nozzle instead of a divergent horn, since one generally wants to excite selectively a particular part of the component. For the sake of the health of the drive units it is important to note that the power handling capacity, quoted by the makers, is correct if the unit is driving a horn which fully loads the diaphragm at the frequency fed to the unit.

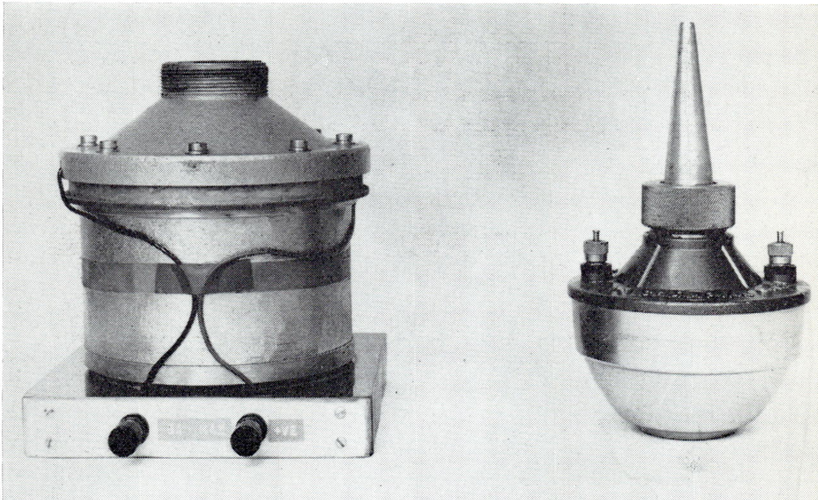

PLATE 40. Two pressure units such as are used commercially for driving horn loudspeakers, the small one is of 12 W rating, the larger one (suitable for hailing large crowds) of 100 W.

If attempts are made to feed the maximum rated power to a unit without horn, or with a convergent nozzle, the acoustic match between diaphragm and air column is poor and the rating has to be considerably reduced at low frequencies, or the moving coil will batter against the stops.

The type of jobs for which this mode of excitation is useful fall broadly into two classes. When checking natural frequencies of high resistance non-magnetic components where it is desired to avoid direct loading of the part, as by a moving coil vibrator, they are often used ; titanium compressor blades, for instance, or nimonic turbine blades, are specific cases in point. They are also invaluable when trying to simulate engine excitation of sheet metal parts, such as hollow inlet guide vanes or other canware, especially at frequencies above about 500 Hz.

A disadvantage of this type of exciter is the high sound levels produced throughout the whole sound room if large powers are used, which may dictate the use of ear defenders by the operator. Nevertheless when, for example, stainless steel parts of thin gauge have to be checked there are few alternatives available which do not load the part unduly.

3.6 *Magnetostrictive drive*

In certain special applications the component may be made into its own drive unit. An example where this is very convenient is when the longitudinal resonances of a steel bar-form component, such as a valve for a piston engine, are required.

If the part is placed on the axis of an air-cored coil, which is driven from an oscillator and amplifier, the length changes produced in the bar will drive it longitudinally, without any physical contact. The method will sometimes work well enough on only slightly magnetic alloys.

It must be remembered that the length changes in the part will be of the same sign, whatever the polarity of the induced magnetism, hence the vibration will be at twice the frequency of the coil drive oscillator.

3.7 *Electrostatic drive*

If an electrode of adequate size is insulated from, but in close physical proximity to, the part a high voltage potential between them will attract or repel one from the other. By this means it is possible to cause the part to vibrate at its resonant frequencies. The method is used, in particular in determining the damping coefficients of materials at high temperatures, *when the operator can be adequately protected from contact with the high voltages involved.* The method is not safe as a drive in normal exploration work, because of the real danger of shock.

4. *Methods of detection*

As indicated several times, in earlier parts of the book, the human ear is a most valuable tool in finding resonant peaks. This is one reason for having a sound room in which to carry out this type of work. However, the ear is not conspicuously good at deciding the actual frequency (except in rare individuals), and it is particularly difficult by ear to be certain between frequencies of 2 : 1 ratio, i.e. an octave. There is almost always need to confirm the frequency, at which the part under test is vibrating, by some other means.

However, the ear assisted by an ordinary medical stethoscope is very suitable for tracing nodes and anti-nodes on large parts at very low frequencies, e.g. below 30 Hz. In some inexplicable manner too it is possible to hear the change of phase as one passes the stethoscope over a node. For such objects as jet pipes and the like, whose geometry is against the use of sand patterns; and which are made of thin sheet, which would be affected by contact with a transducer, the stethoscope is quick, convenient and accurate.

Another rather general method used in the frequency and modal pattern checking of vibrating objects—especially at low frequencies—is an optical stroboscope. There are, however, certain vital precautions which must be taken. Stroboscopes are now readily available with high light outputs, but there is a particular medical phenomenon which must be guarded against and also a mechanical ambiguity which can upset frequency checks. Neither of these is as widely known as it should be.

In susceptible persons a flickering light (in the region of the frequency of the alpha rhythm of the brain) can produce epileptic precursors which sometimes produce a fit. The same conditions can also cause emotional and/or hypnotic changes in people who are not potentially epileptic. The stroboscope is used, in medicine, as a trigger in the investigation of suspected epilepsy. Many readers may be aware of a similar effect in connection with high-powered ' pop ' music, associated with powerful flickering lights in dance halls. The critical frequency lies between 5 Hz and 15 Hz (300 to 900 r.p.m.).

Therefore, when using stroboscopes for viewing vibrating objects *no one should ever be alone*; particularly in a sound room where people outside cannot see what is happening. Operators should be screened, by a medical practitioner, before doing work with light frequencies in this particular band—or near it.

It is, in any case, easy to avoid the critical band of frequency, as far as the light is concerned (which is what matters) by working the stroboscope at one half, one third, etc. of the frequency of a part which is vibrating in the critical range. This will give an equally sharp, though not so bright, an image.

As far as ambiguity is concerned, the stroboscope will give a clear image when its rate of flicker is any submultiple of that of the vibrating part as noted above, i.e. it is as easy to view the component every alternate cycle or even every three or four cycles. Hence it is easy to make a mistake in the frequency unless other means, as detailed later on in this chapter, are adopted to confirm the correct frequency. Without such cross check the highest strobe frequency which arrests the motion is correct—if one is sure the strobe will tune that far! An example of a powerful stroboscope for sound room or engine work is illustrated in plate 41.

Another general area in which precautions are necessary, as they would not be in engine or rig test, is ' crosstalk ' or mutual interaction between excitation and detection. There is a need, as described in earlier chapters, in engine and rig testing to screen the signal wiring of measurement devices to avoid spurious frequencies of interference. However, these unwanted frequencies, while they confuse or blot out results at times, rarely do so at the same frequency as the genuine signal (except, as noted, when wiring loops generate a 1 EO signal which coincides with the effect being measured).

In the sound room many of the methods of excitation used are basically fed from a power source of an electrical nature, whose frequency is the same as that of the vibration of the part which is being measured. This implies that special care must be taken to guard against pickup, of an electromagnetic or electrostatic nature, by the detector from the exciter when working with electrical excitation and detection.

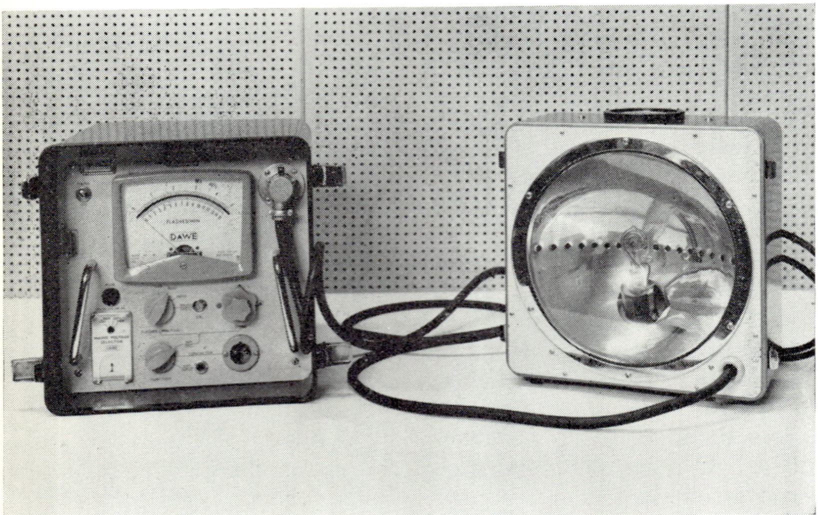

PLATE 41. A very powerful stroboscope suitable for sound room and engine use, but requiring care because of its large output.

The effects of electromagnetic induction are worst at low frequencies, and of electrostatic (or capacitative) transfer at high frequencies. Often the easiest way to deal with these effects is to take care to keep the two sets of wiring physically well separated. For example when exciting a turbine disc with any magnetic exciter at one point on the periphery, the detector probe should be placed over a diametrically opposite point and leads should all be screened and input and output kept well away from each other.

Another way in which the problem can be solved very neatly is to use a frequency modulated detector or use amplitude modulation on a carrier frequency well removed from all test frequencies. There is no trouble, of course, when mechanically exciting the part from a violin bow, nor when electrical excitation is combined with acoustic detection, as by a microphone some feet away from the exciter, or stethoscope and the ear.

It is when the excitation is electrical and the detection is by accelerometer, seismic pickup, direct capacity pickup, straingauge or piezoelectric device that most care must be exercised. If such interference is suspected, just stop the part with the hand, if possible, or clamp it mechanically; and crosstalk will remain while genuine signal will disappear.

4.1. *Nodal pattern detection*

Various references have been made to the detection of nodal lines so that it is not necessary to expand unduly in this section. Some information on the use of sand patterns, holography, etc. appears worth while. If dry sand is sprinkled on a horizontal flat surface which is vibrating, the sand will move with the surface until a vertical acceleration in excess of 1 *g* occurs. The result of this is that when, say, a disc is vibrating by say 5 *g*, or more, at the antinodes, there will be a line each side of each antinode where just 1 *g* will be felt. Sand nearer the antinode than this will be thrown out of contact—preferentially towards the lower acceleration side—so that, after a short while, the sand will be moved from all areas experiencing more than 1 *g*. If fairly large accelerations occur (particularly likely the reader will remember at high frequencies) then these double nodal lines will coalesce; resulting in the familiar sand patterns illustrated here and earlier in the book. Where applicable sand patterns are just about the quickest of all methods.

A stethoscope can be used to advantage at lower frequencies, say from 2 kHz down, and is particularly useful when dealing with large objects, such as jet pipes and the like; with this method surfaces can be vertical or upsidedown.

Traversing the other detectors, described later, over the surface of the part can deal with other cases; but this takes longer and care has to be taken, not only to avoid crosstalk between leads but to make sure the contact types of devices, e.g. accelerometers, do not damp down the resonance, or even shift the node.

Undoubtedly the most sensitive method—and one where no interference with the vibrating surface can be caused—is the use of holography. While rather expensive in the first case, holography conveys so much information as to the deflection state of the vibrating object that, in the author's view, it will gradually outdate most present methods of nodal pattern analysis for small and medium sized objects. A bonus of considerable benefit arises from the fact that very small deflections are needed and a 12 W air pressure horn unit will produce usable deflections on a 0·25 in. thick plate, if the *Q* is reasonably high.

Where holography really comes into its own, however, is when the movements of a complex assembly, e.g. a two or three-disc turbine assembly with blades, has to be determined. The *modus operandi* of holographic methods is too involved to explain in the present book, but the references in the bibliography will give full explanations. At the time of writing there is no standard commercially available equipment

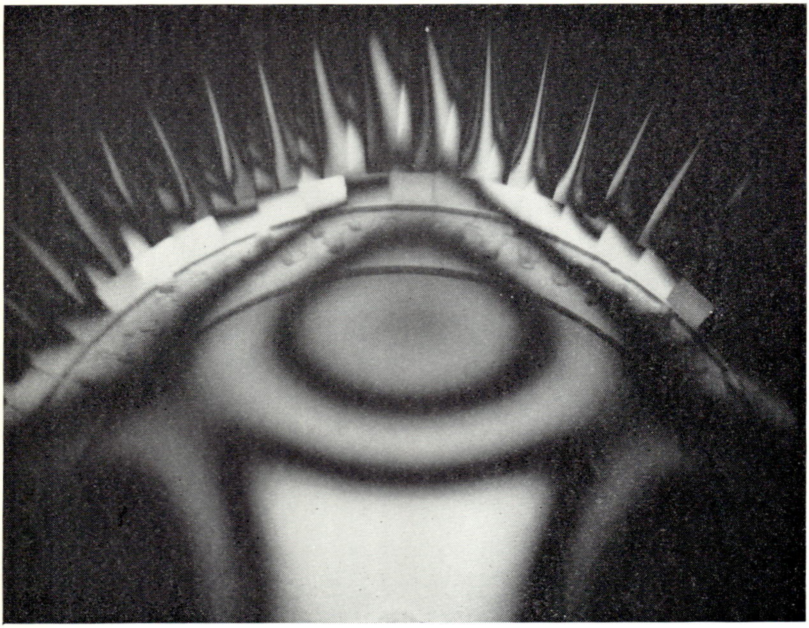

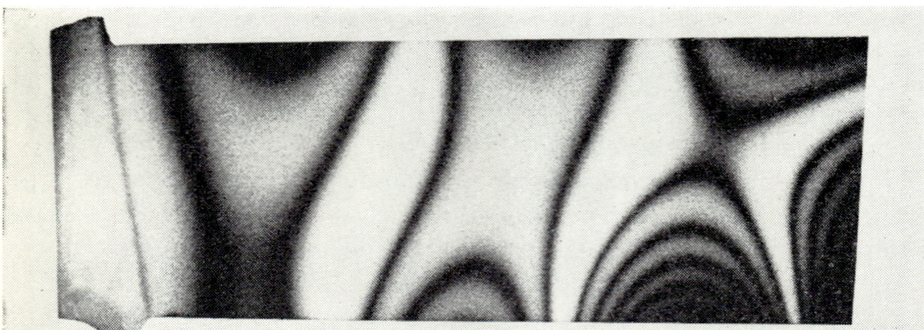

PLATE 42. Two time-averaged holograms. The top one is of part of a bladed turbine disc, note how the vibration of the blade and disc couple together. The lower one is of a compressor blade excited by a peizoelectric crystal fixed to the surface.

as a complete unit. Plate 42 shows typical time average holograms of vibrating parts.

4.2. *Straingauge calibrations*

An important feature of sound room work is calibrating straingauged components dynamically. From reasons of economy in time of preparation for test or because of limitations of channel availability; or to reduce the effect of gauges on the part itself, it is often necessary to use gauges to read the stresses in more than one mode of the component. Where this is the case the gauge cannot be placed in the optimum position for all modes. It is, however, possible to carry out sound room tests, exciting the component in a variety of likely modes and obtain factors for the maximum stress related to that read by the gauge. In the sound room test more than one gauge may be employed.

Also, when attempting to relate stress read in the engine or rig, to that obtained in fatigue tests in particular modes it is often convenient to compare amplitude with straingauge output and use, e.g. tip amplitude as a parameter in the fatigue testing. This avoids straingauging each fatigue specimen—particularly desirable when the fatigue is performed by air-blown self-excited methods. In this case the same size and position of gauge is used for engine and sound room test.

In sound room tests we know the frequency we want and filtering (by using a selective amplifier) of the straingauge signal is wholly permissible ; and indeed often necessary because the excitation power available is so much lower than that operating in the engine, and quite low stresses will be experienced.

Sound room work is particularly satisfying for the vibration engineer because he is—just for a while—removed from the stresses and pressures, both technical and political, of live engine and rig testing. There need be no one breathing down his neck (as long as he locks the door) so that tests can be repeated at will. A variety of methods can be employed to crosscheck each other, so that the results can be satisfyingly accurate—frequency in particular can easily be correct to 1 part in 10,000.

Contrary to some customers' impression the vibration engineer has a normal conscience, and prefers to obtain accurate results; though the conditions within which he has to work often make this wellnigh impossible. Small wonder then if he regards his sound room as a haven of refuge, a sanctuary from the wicked world of the project or production manager who not only wants the results the moment the engine is rigged on the test stand, if not sooner; but wants them checked and neatly set out in graphs to boot!

BIBLIOGRAPHY

RALEIGH, LORD	*The Theory of Sound.* Reprint, 2 vols.	Dover Publications, 1945
DEN HARTOG, J. P.	*Mechanical Vibrations*	McGraw Hill, 1947
TIMOSHENKO, S.	*Vibration Problems in Engineering*	Van Nostrand, 1955
MORSE, P. M.	*Vibration and Sound*	McGraw-Hill, 1953
FLETCHER, H.	*Speech and Hearing*	McGraw-Hill, 1953
RICHARDS, E. J. and MEAD, D. J.	*Noise and Acoustic Fatigue in Aeronautics*	Wiley & Sons, 1968
Wood, A.	*Physics of Music*	Methuen, 1964
YARNELL, J.	*Resistance straingauges*	*Electronic Engineering*, 1951
NEUBERT, H. K. P.	*Straingauges: Kinds and Uses*	Macmillan, 1967
PATMA, T.	*Straingauges Theory and Application*	Iliffe Books, 1967
HOCKLEY and BUTTERS	Holography in vibration analysis	*J. Mech. Sci.* **12**, No. 1, 1970
JOHNSON, D. C.	Free vibration of an elastic body	*Aircraft Engineering*, **XXIV**, 1952

For names of makers of transducers and equipment and details of straingauge configurations, sizes and properties consult *Instruments : Electronics : Automation : Year Book* published by Morgan Bros., and write for catalogues.

For interesting reading on the construction, playing and acoustics of musical instruments read Ernest Benn's series : " Instruments of the Orchestra "—*Clarinet, Oboe, French Horn, Trumpet and Trombone, Bassoon and Contra Bassoon.*

GLOSSARY

Note : When preceded by ' In context ' the definition is appropriate to the use of the term in this book.

Analogue/Digital Converter : An electronic device which converts a continuously varying quantity into a quantized representation in decimal, binary or other code.

Analogue Output : A device which follows continuously the variation of a given parameter is said to have an analogue output.

Anti-resonance : That frequency, of an oscillating circuit or mechanical vibrating system, at which it presents the maximum resistance to excitation.

Aperiodic : In context ; a vibration, or excitation, which is random, non-current or non-cyclic. An aperiodic excitation can cause a periodic vibration at a part's natural frequency.

Axial Vibration : In context ; the vibration of an engine or part in the direction of the engine axis, i.e. that of the main shaft(s).

Band Pass Filter : An electrical, or mechanical, circuit combining a low-pass filter with a high-pass filter so as to pass a restricted band of frequencies with uniform response and attenuate frequencies sharply on either side.

Beats : The phenomenon produced when two different frequencies act on the same part, if these are near in frequency. A waxing and waning results as the vibrations alternatively reinforce and cancel. It implies a non-linear response by the part.

Bi-morph Crystal : A sandwich of piezoelectric material—natural or artificial—so arranged that it responds primarily to bending.

Bourdon Effect : The effect of internal pressure which tends to straighten a bent tube if oval in section. Is used in the simple Bourdon pressure gauge and other transducers.

Circular Gauges : Straingauges wound in a flat spiral. They are used when it is desired to take an output from a pressure diaphragm in a transducer.

Cross-sensitivity: The response of a transducer to an input of the right sort in the wrong direction. E.g. the response of a vibration pickup to movement at right angles to its sensitive axis.

Cyclic Vibration : A cyclic vibration is one where the pattern of deflection recurs regularly—even though the waveform is not itself simple.

Dashpot : A mechanical viscous damper, where a piston attached to the part to be damped moves in a chamber so as to pump gas or liquid against a resistance.

Digital/Analogue Converter : An electronic device which converts a digital input, such as a series of pulses, into a continuously varying voltage proportional to their numerical value.

Digital Output : A device which quantizes the response to a stimulus. This may be by converting to a decimal display, or to a binary or other code.

Doppler Effect : The effect whereby wave motion has its frequency changed when reflected back from an object moving with respect to the observer.

Earth Loops: The colloquial term given to the effect of multiple earth paths or leakages between items of equipment, which results in a.c. potentials being developed which are presented to the readout equipment as genuine signals.

Fibreoptic Device : An arrangement of many small glass filaments in a bundle, so that light falling on one end will be transmitted to the other. According to the method of manufacture the spatial picture may be preserved, or just the general level of illumination.

Forced Vibration : A vibration, determined almost entirely by the force acting on the part.

Gunn Diode : A semiconductor diode which exhibits a negative resistance characteristic which makes it suitable for use as a very small microwave oscillator.

H.P. ; In context ; the high pressure part of an engine, in particular the compressor or turbine associated on the same shaft of a gas turbine engine, which deals with the highest stages of compression and expansion.

High-pass Filter : An electrical, or mechanical, circuit which presents a uniform response over a changing range of frequencies until a certain lower value is reached, after which the response decreases sharply.

Holography : In context ; holography is a method by which both the phase and amplitude of the wavefront scattered from an object illuminated by coherent light are recorded in the form of an interference pattern on a photographic plate (see Bibliography, Hockley and Butters).

Interference Electrical : Basically is unwanted electrical impulses, not germane to the quantity being measured. Analogous to noise present when speech or music is wanted.

L.P. : In context; the low-pressure part of an engine, either compressor or turbine in the case of a multi-spool gas turbine.

Low-pass filter : An electrical, or mechanical, circuit which has uniform response over a range of frequencies until a certain higher value is reached. After this point the response decreases sharply.

Magnetically Hard : Parts of ferromagnetic material of such properties that they have a high coercive force, i.e. once magnetized they retain a substantial proportion of the magnetism when the field is removed. Most hardened steels come into this category.

Matrix Methods : In context; separating the part into elemental sections and using matrix algebra to solve the resulting simultaneous equations.

Natural Frequency : When an elastic body is deformed and the deformation released it vibrates at a natural frequency in one or other of its possible modes. In practical terms is indistinguishable from resonant frequency.

Photovoltaic Effect : The property of certain semiconductors to develop an electromotive force when light falls on them.

Piezoelectric Effect : The property of certain natural crystals, such as quartz and Rochelle salt and also some semiconductors, whereby when the crystal is strained a charge appears on opposite faces of the crystal.

Pitot-static Tubes : An arrangement where one tube faces into a moving air stream, while another faces away. When coupled to a differential pressure transducer the device registers a difference proportional (but not directly) to the velocity of the air.

Probe : In context; a device attached to a transducer so as to conduct the effect being detected before transduction. In the case of a microphone a simple tube; with a photoelectric sensor a fibreoptic light pipe; with a displacement transducer a coupling rod. Used when space or adverse ambient conditions preclude the transducer being situated where the measurement is required.

Radial Modes : In a large-diameter pipe containing gas, that mode with a displacement node along the axis and another round the periphery.

Random Vibration : A vibration which has no cyclic basis, or consists of many discrete vibrations randomly combined, which result in non-repetitive motion.

Resonant Frequency : Vibration of a component, when the exciting force approximates one of the natural frequencies; under these circumstances that frequency where maximum response occurs is the resonant frequency. In practice it is indistinguishable from the natural frequency.

Rocking Vibration : In context; a torsional vibration normal to the engine axis. The term is used because most engineers regard " torsional vibration " as inevitably about the engine axis, hence rocking vibration is less confusing.

Roll Off–in Filters : An electrical filter which is frequency sensitive usually shows no change in response until a certain frequency is reached, when the characteristics change sharply (but not suddenly). The frequency at which the response deviates from frequency linearity is called the roll off frequency.

Rosette Gauge : Separate or assembled straingauges fixed to a surface so that their active axes intersect with the gauges in different angular directions. The use of such an arrangement permits the calculation of the surface strain in any direction.

Rotating Stall : In a compressor a phenomenon preceding full stall when locally stalled patches of air travel round one, or more, stages of blading in an axial plane. Always occurs at less than the speed of the rotor and can cause resonance with the blades.

Spectrum Analysis : In context; an analysis of the response in terms of its individual frequency components; by analogy with the separation of light into separate colours.

' *Static* ' *Electric Effect :* In context; the effect of an ionized gas flow past a component, which can transfer randomly the charges from the ions to the part. This causes spurious voltages to appear as noise in associated circuits.

Swash Plate Pumps : Plunger pumps where the reciprocating motion is derived by rotating a body with several spring-loaded pistons against a ' swashed ' or angled camplate.

Tangential Vibration : In context; the vibration, of a torsional nature, at the periphery of an engine, the motion being tangential to a circle whose centre lies on the engine axis.

Telemetry : In context; obtaining in a stationary receiver, by modulation of electromagnetic waves, information from a moving part.

Time Average : The term used when a signal is averaged out over a time many times greater than the time of oscillation. The value so obtained does not reflect the maximum value nor indicate the time composition of the parameter involved.

Torsional Vibration : Angular vibration about an axis—generally that of a shaft or an engine carcass.

Transducer : Any device which changes one physical effect into another. The human eye changes light into electrical impulses (in coded form), a straingauge changes strain into resistance effects and so on.

U/V Galvanometer : In context; a D'Arsonval or ' string '-type galvanometer, illuminated by a high-pressure mercury vapour lamp.

Generally, in vibration work made in the form of a pencil-shaped unit, several of which may fit into a common magnetic block and thus form a multichannel recording unit.

Wake : In context ; the disturbance, either up or down stream, in the airflow caused by a strut or by asymmetry of an engine intake or similar duct. Wakes produce strong resonance effects when rotating blades pass through them at each revolution.

Waveform : The plot of the variation of a cyclic vibration against time—as displayed on a cathode-ray oscilloscope or on moving film or paper. Waveforms can vary from a simple sine wave (plot of a simple harmonic motion) to very complicated ones when many different effects combine in causing the vibration.

Wiring Loops : In context ; if the wiring to and from a straingauge or other transducer rotates in a magnetic field (even as weak as that of the earth) electric voltages are induced—as in a dynamo coil. Since these are strictly related to the frequency of engine rotation they are often mistaken for genuine signals. The effect is the larger the greater the area enclosed by the wiring loop and the higher the field.

INDEX

The index was generously prepared by Mr. R. Graham, a recent colleague of Mr. Wallace.